REFLECTIONS
From Earth Orbit

Winston E. Scott
Captain, USN, Ret.
NASA Astronaut

The STS-87 mission crew, during Terminal Countdown Demonstration Test activities at Kennedy Space Center, pose in front of the space shuttle *Columbia* being prepared on Pad 39B.

REFLECTIONS
From Earth Orbit

Winston E. Scott
Captain, USN, Ret.
NASA Astronaut

Dedication

To my wife Marilyn Scott; Son, Navy Lieutenant Winston Scott; and daughter, journalist Megan Scott.

In memory of my mother Rubye Scott who, for my space flights, had the best seat in the house.

To my personal friends and professional colleagues who perished on board *Columbia*'s STS-107.

All rights reserved under article two of the Berne Copyright Convention (1971).
Published by Collector's Guide Publishing Inc.,
Burlington, Ontario, Canada, L7R 1Y9
Printed and bound in the USA
REFLECTIONS FROM EARTH ORBIT
by Winston Scott
ISBN 1-894959-22-1
ISSN 1496-6921
Apogee Books Space Series No. 52
©2005 Apogee Books

Acknowledgments

No one can accomplish anything of significance without the help and encouragement of others. Writing this book, a significant accomplishment for me, would not have been possible without the support of the following persons.

My wife Marilyn was always the first to read each chapter of my manuscript and provide feedback. Her perspective was invaluable to me in my attempts to take a complex subject, space flight, and make it understandable, interesting, and relevant to the average reader. Also, her expert computer skills allowed retrieval of more than one lost paragraph from cyberspace.

My thanks goes out to Dr. Kathleen Colgan, who generously contributed her time to reading and editing my drafts.

Thanks also to Dr. Pat Cooke. Pat and I played trumpets with the Tallahassee Swing Band. Pat would use his break time between sets to review my manuscript with me. Of all, Pat Cooke provided the most positive feedback and encouragement.

I express my appreciation to retired naval aviators Captains Buddie Penn and Chuck Nesby. Chuck and Buddie provided support and encouragement throughout my twenty-seven year navy career. I could not have dreamt of better role models and friends than these two superb officers and talented aviators.

Thanks to my in-laws (the late) Albert and Josephine Robinson. The support and encouragement they gave to me is greatly appreciated.

Last, but not least, I an indebted to my family. My father Alston, mother (the late) Rubye, brother James, and sister LaVerne. Together we were as close as possible to being the perfect family unit. It is my sincere hope that my family members will read this book, reflect on my life, and realize that their patience, so freely provided to me, has been well worth the investment.

Contents

1. Where Have You Gone Sky King 7
2. I Never Called Him Uncle Willie 23
3. And What May Come of Dreams – Ascent ... 39
4. Day Pass 53
5. A Model Plane For Winston 60
6. How Could They Know 67
7. The Best Seat In The House 74
8. Night Pass 85
9. We've Got To Have Arithmetic 90
10. It's About Time 99
11. To Walk Outside Columbia 106
    Biography 127

— Chapter 1 —

Where Have You Gone Sky King?

"Out of the clear blue of the western sky, comes Sky King!"

THAT WAS THE FAMILIAR AND EXCITING LINE THAT BOOMED from the old black and white TV into our 1950's living room each Saturday morning. I must have been about eight years old at the time, and my brother James (we called him "Junior" because he was named after my father) must have been six. Saturdays were fun-filled days in our home. Even though there was no school, we awoke early. With jet-like speed, we would jump out of bed, wash our faces, brush our teeth, comb our hair and get dressed. Then, fueled by the super energy from that cereal of champions, we'd set land speed records as we made our beds, cleaned our room, emptied the trash, and performed what, to two young boys, appeared a seemingly endless list of other chores. But once those chores were done, we were mom-approved – and ready – for the regular lineup of Saturday morning TV shows. And what action-packed shows they were! There was the original *Rin Tin Tin*, *The Tales of Texas Rangers*, and the usual lineup of cartoons, such as *Mighty Mouse* and *Popeye*. Despite the brutal competition, our all-time favorite show was *Sky King*.

Schuyler J. "Sky" King was a flying cowboy who, on a weekly basis, thwarted the plans of some of Hollywood's most dastardly villains. While most TV heroes of the time rode horses, Sky King flew an airplane called the *Songbird*. Our flying hero's sidekicks were his nephew Clipper and his niece Penny, both of whom could fly the *Songbird* almost as well as "Uncle Sky."

Like most kid's TV shows of today, the weekly plots of these Saturday morning shows were predictable. But to a whole generation of young children (and especially boys like Junior and I), these shows were exciting, and their plots were intricately suspenseful and complex. And we knew those plots by heart. The villains, having

In the early shows of the series, Sky King flew a Cessna T-50, similar to the one above, which he had named the *Songbird*. For most of the series, however, he flew the *Songbird II*, a Cessna 310B like the one shown below.

committed one crime or another, would be making a fast get-away, the sheriff in hot pursuit. Even in the TV world, where good always triumphs, the sheriff most often ran the risk of losing the bad guys, who always seemed to have an unfair head start. Attempting to escape justice, the criminals would be driving at breakneck speed, over hills and through canyons, with dust filling the air behind their get-away car. Swerving from side to side, their car would narrowly miss boulders and cliff edges. Then, from over the horizon, out of the blue, came Sky King! Flying the *Songbird* at maximum velocity, Sky King would swoop low over the get-away car, criss-crossing back and forth, causing the crooks to lose control of their vehicle and run off of the road. He would then land quickly, run to the disabled get-away car, and attempt to apprehend the villains. Of course, a fight would almost certainly begin; but not to worry – Sky King could fight better than any two or three (or four hundred) bad guys. He would quickly win the fight, capturing the villains just as the sheriff arrived on the scene to take them into custody.

Ah-h-h, the life of Sky King!

And what a great life it was to two little boys in the 50's.

Although I enjoyed watching Sky King fly his aircraft, I never gave much thought to the possibility that, some eighteen years later, I too would be a flyer.

It was September of 1973 when I began Navy flight training at the Naval Air Station (NAS), at Saufley Field in Pensacola, Florida. I had recently graduated from Florida State University with a degree in Music. Unable to explain my strong and mysterious desire to fly, I gave in to the urge and entered the Navy's Aviation Officer Candidate School (AOCS) at NAS, Pensacola. The public got a Hollywood glimpse of AOCS in the 1980's movie success *An Officer And A Gentleman*, which starred actor Richard Gere.

Physically, mentally and psychologically demanding, AOCS was by far "no picnic," a perception the movie did a fair job of showing. AOCS lasted about four months. Each day during those four months, we awakened at 0500 (5 a.m. for you civilians) to Reveille. Within five minutes we had to be dressed and in formation for physical training. Push-ups, sit-ups and squat-thrusts were preliminary activities to running. And we were most definitely "off and running." Every minute of the rest of the day was filled with one activity or another. There was military drill during which we would march in formation wearing fatigues, boots and helmets, carrying our "pieces" (M-1 Carbine rifles). There were academic classes in military history, military etiquette, leadership, aerodynamics, engineering, meteorology, public speaking, and well, you name it! There was swim training which, after several weeks, culminated in a mile-long swim – wearing flight suit, boots and helmet. There was water survival in the Atlantic Ocean, land survival in a nearby forest, and training to escape from a crashed and submerged aircraft. This went on day after day, week after week, and month after month. Each candidate had to successfully complete AOCS before being permitted to even get close to an airplane.

> **They could kick me out, but I was absolutely *not* going to quit.**

Several AOC's who had begun at the same time that I did had already either washed out ("attrited") or quit (been "dropped on request" or DOR). But I was determined to succeed – at least, I was determined not to give up. They could kick me out, but I was absolutely *not* going to quit. At the successful completion of AOCS, I was finally ready for my first actual flight lesson in an airplane.

On the day of my PS-1 (pre-solo-1) flight, the first flight of the curriculum, I arrived an hour or so early at the squadron ready room, where all pilots meet to prepare for their scheduled flights of the day. I was confident and eager. Even though I had never before been in the cockpit of an airplane, I knew – and I knew without a doubt – that I had the potential to fly and fly well.

My flight instructor met me in the ready room and introduced himself as First Lieutenant Peterson, Marine. He was of moderate stature, about 5 feet, 10 inches, and maybe 165 pounds. With a trim physique and a firm handshake, he was in possession of the standard-issue Marine Corps close-cropped haircut. His demeanor, like those of all primary flight instructors, was curt and very much to the point. Yet, despite his military demeanor and his snappish and systematic efficiency, he was considerably more personable than the Marine Corps drill instructors that I had encountered in AOCS. First Lieutenant Peterson was only a few years older than I was, but he possessed something that I did not yet have – those navy wings of gold, proudly pinned to his chest. I was confident that, one day, I too would wear gold wings.

During PS-1, the instructor demonstrated the preflight activities, such as donning of the flight equipment and inspection of the aircraft. He also demonstrated all of the flight maneuvers to be performed in the aircraft. Although this was my first flight, I was responsible for having studied and thoroughly prepared myself for the flight. I knew what Peterson would be looking for. He would try to determine how well I had studied and practiced the preflight inspection sequence, a sequence during which we would examine the condition of the aircraft to assure its suitability and safety for flying. But preflight inspection was merely the beginning of this exercise. He was also intent on ensuring that I had studied and understood the aircraft engine pre-start, start, and post-start procedures. Furthermore, I was supposed to understand the purpose and function of every lever, gauge and dial in the cockpit of the aircraft. I was also required to understand how to interpret or use each and every one of these switches, levers, gauges, and dials. Additionally, I was required to know the written procedures for the in-flight maneuvers, and for selected emergency procedures.

As important as these and other operational aspects of flight were, there was something else Peterson would be grading, and that grade would be the most important grade I would receive all day. It could determine whether or not I would indeed, one day, wear wings of gold. This all-important grade would spring from Peterson's assessment of my aeronautical adaptability. In other words, based on my performance, he would try to determine how well I fit into the airborne world of military aviation.

Peterson was responsible for making subjective decisions about my confidence in the cockpit, and my ability to deal with flight situations in a level-headed and clear-thinking manner, and my aggressiveness in responding to potentially dangerous events. From take-off roll to landing roll-out, he would be monitoring my apparent overall psychological comfort levels in the flight environment. In particular, Peterson would also be watching for signs of what was one of the greatest concerns to all new military flight students, an issue which obsessively worried prospective naval aviators who have never before flown an airplane.

The concern? Quite frankly ... *airsickness*!

Why was airsickness such a profound concern for aspiring pilots?

To become airsick is to be uncomfortable and embarrassed. To become airsick is a proclamation to the world that you don't have "the right stuff." Chronic airsickness earns a student pilot a one-way ticket home, and away from the world of naval aviation.

A T-34B Mentor trainer aircraft at the Naval Air Technical Training Center (NATTC) on board Naval Air Station Pensacola, Florida. (U.S. Navy photo.)

The aircraft used in Navy primary flight training was the T-34B Mentor, the military version of the civilian Beechcraft Bonanza. The T-34 was single-engine, low-winged, and propeller-driven, with fore and aft (tandem) seating. Like all Navy training airplanes, it was painted a bright "international orange" and white, so that student pilots could more easily see other training airplanes, and be seen by other training pilots. As the instructor, First Lieutenant Peterson sat in the rear cockpit while I, as the student pilot, sat up front.

Prior to walking out to our aircraft, we obtained our parachutes from the paraloft and inspected them. After placing our parachutes in our respective cockpits, we began our preflight inspection. Pointing to each aircraft part and talking about its function, First Lieutenant Peterson completed each step of the preflight checklist. He explained the condition in which he expected to find each aircraft component. In his own clipped tone and blunt way, he occasionally asked me a question about a particular aircraft part and its function. I had studied well and answered most of his questions.

At one point during the preflight inspection, Peterson, with a wry "gotcha" smile, pointed to a device attached to the nose wheel of our airplane. "Do you know what this is?" he asked. Proud of myself, I answered, "Yes sir! That's the shimmy

The T-34B Mentor is the military version of the civilian Beechcraft Bonanza, as shown above.

damper. It's a pneumatic device that reduces nose wheel vibrations during high-speed ground operations. It prevents us from damaging the nose wheel mechanism."

> **I ought to drop-kick this dude!**

He stared at me for a moment, and then said bluntly, "It'll keep you from busting your ass!"

My expanded chest deflated and, suddenly dejected, I tried once again to mentally pump myself up . . . *"How dare he not acknowledge what in fact was a brilliant response, a clear demonstration of superior aeronautical knowledge and engineering insight. I ought to drop-kick this dude"!*

However, I opted for "the better part of valor" and simply responded, "Yes sir."

We completed the inspection looking at every inch of that airplane, front to rear, top to bottom, and port (left) to starboard (right). No inspection in the history of aviation (or microsurgery, for that matter) had ever been accomplished in a more thorough manner. Funny thing, though, I don't recall Sky King ever doing a preflight inspection.

With the preflight inspection finally complete, we climbed into our T-34. We secured our seat belts and shoulder harnesses, put on our helmets, and put our microphones in place. First Lieutenant Peterson turned on the battery switch. The aircraft suddenly came alive, making humming and whirring sounds as if eager to have its engine started, so that it could leap off the ground. Peterson and I, each in turn, depressed our respective throttle-mounted buttons to key the Interphone Communications System (ICS). The T-34 cockpit is relatively noisy with the engine running, and we would certainly need the ICS if we were to talk to each other during the flight.

I tried as best I could to watch and understand as First Lieutenant Peterson performed each step on the pre-start checklist. I then watched and listened as he started the engine and performed the post-start checks. These checks are meant to ensure that the engine, flight controls, and all pieces of aircraft equipment, are operating properly. As we began to taxi out of our parking spot, Peterson keyed his ICS and began to recite what sounded like a well-rehearsed monologue.

"Taxi is the controlled movement of the aircraft along the ground. Speed is controlled by power and direction is controlled by rudder, augmented as necessary by the brakes." Peterson's tone was so practiced and perfunctory that, at first, I wasn't sure whether it was him talking in real time, or a recording, like one of those instructional recordings that the airlines sometimes use to explain the safety and emergency procedures to their passengers.

We satisfactorily performed engine checks and received take-off clearance from the control tower. After we had positioned ourselves on the runway for take-off, First Lieutenant Peterson added power, checked the flight controls once more, and then released the brakes. The little T-34 lurched forward and quickly gained speed. I could hear the wind noise getting louder and louder through the open canopy. I could smell the exhaust of burned Avgas (aviation gasoline). As we passed through approximately 55 knots (roughly 60 miles per hour), Peterson rotated the nose upward and, within moments, we were airborne.

I felt excited and, as I had always anticipated, confident and secure in this new world of military flying.

I'm coming to join you Sky King!

We continued to climb, gaining altitude, and within a few minutes left the airport traffic area. We slammed our canopies shut and turned towards the practice area. It took us only about ten minutes or so to arrive, but once inside the practice area, First Lieutenant Peterson began to demonstrate some of our basic flight maneuvers.

First, there were a few level speed changes, then a normal climb, followed by normal cruise, a series of constant-altitude turns, and finally, slow flight. As he executed each maneuver, he sharply recited the procedures I was shortly to imitate. In his now familiar and no-nonsense manner, he said, "a normal climb is performed at 100 knots. Set mixture, rich; prop (propeller), full increase rpm; throttle, full open. Raise the nose to initiate the climb, and then adjust to maintain 100 knots. Trim the aircraft."

Since these were very basic maneuvers, he let me try each one on my own, giving me practice in maneuvering the aircraft and manipulating the engine controls. It also gave him an opportunity to observe how well I had memorized, and could apply, my flight procedures.

The Florida sun was bright, and it shone relentlessly hot through the top of the magnifying glass that was our aircraft canopy. With no air conditioning in the aircraft, the cockpit was uncomfortably warm. Furthermore, what air there was in the cockpit reeked of engine exhaust.

> I was, as the old Navy saying goes "fat, dumb, and happy."

Still, as I maneuvered this navy aircraft through the Florida skies, high above the ground, I felt pretty good!

We repeated these basic maneuvers, and then conducted an area-familiarization sequence, during which Peterson pointed out the various landmarks and checkpoints that I could use to keep myself from becoming lost during my solo flights in Pensacola airspace. I continued to fly the aircraft, except on occasion, when Peterson briefly took the controls. When he did so, I pulled out my map from my flight suit pocket and tried to locate on the ground the various landmarks on its face.

I was pleased. So far, it had been a good flight. There I was, with my hands on the flight controls, in a bona fide Navy aircraft. I was, as the old Navy saying goes "fat, dumb, and happy." I could not have anticipated the impact the next string of events would have on my first experience flying an airplane.

Peterson suddenly took the aircraft controls from me, keyed the ICS, and boldly announced, "Hold on, here we go." Rolling the aircraft left, he did a complete 360-degree flip. A moment later he asked "How do you like that?"

I truthfully answered, "That was fun. Can we do another one?"

He rolled right this time, again wings turning, completing 360 degrees of movement. I again keyed my ICS and declared, " That was really fun, let's do it again."

This time, the seasoned aviator added power and pushed forward on the control stick. The aircraft went nose down and began to pick up speed. Instinctively, I tightened my stomach muscles against the negative g-forces that wanted to push everything inside my body upward (particularly the inside of my stomach). I briefly began to float, but my seat belt held me close to the seat cushion. Peterson then pulled back on the control stick and the nose of the aircraft began rotating upward. I could feel the positive g-forces increase as the aircraft climbed. This time my insides wanted to move downward as the g-forces pressed me further and further into my seat.

Peterson continued to pull us upward while simultaneously rolling the aircraft towards the left. Now the airspeed was slower as he rolled over onto the left wing. As its nose began to fall through the horizon, the aircraft picked up speed again. And again, I tightened my stomach. As the nose of the aircraft dipped well below the horizon, with its speed dramatically increasing, I could feel the increased g-forces push powerfully against my body, squeezing me further and further down into my seat. The forces on my body subsided as Peterson began to bring the aircraft towards level flight.

For the third time now, I tightened my stomach muscles trying to minimize the discomfort I was beginning to feel.

Peterson keyed the ICS and asked, "How did you like that?"

I answered, "That was great; let's do some more."

In the back of my mind, however, I thought, *I'm starting to feel a little uncomfortable in my stomach. No big deal.*

That Marine pilot in the rear cockpit again added power and lowered the nose. We picked up even more speed than we had on the previous maneuver! As we dove towards the Earth, our aircraft speed and noise building, the cockpit seemed even hotter, and little beads of sweat began to run down my face. I tightened my stomach once again.

Now the plane was nosing up . . . there's a roll . . . suddenly we were completely upside down.

Actually, being upside down in the speeding plane didn't bother me the least bit. What did bother me was this constant up and down, from negative g-forces to positive g-forces, to negative g-forces again – all of which made my stomach churn. These new sensations to my uninitiated body were certainly getting my attention!

After we completed what I recognized to be a barrel roll, Peterson began talking to me again. "What do you think?" he asked. Determined not to let on that I was feeling sick, I lied, "Great! Let's do some more." But what I said did not reflect what I was thinking. I found myself seriously wondering, *Does a person have to be some sort of sadist to be a Naval Aviator? I hope he knocks this stuff off soon. I don't feel so good. In fact, my stomach feels as though I'm going to . . .*

Peterson fiendishly performed another barrel roll, this time in the opposite direction. He again he asked me how I felt, and once again I lied.

Suddenly, it hit me! I knew exactly what Peterson was trying to do. And now I was really beginning to worry. He wanted to see whether or not I had the strength, the gumption, the "cajones" to be a Navy Pilot. He wanted to see whether or not I did indeed have *The Right Stuff*!

I was bound and determined to hold on. He was not going to make me airsick.

Feeling completely miserable, I suddenly remembered something one of my flight school classmates had told me after his first flight. AOC Osterland was a big guy from Carlsbad, New Mexico. Standing about six foot four or five, and weighing a good 250 pounds, he looked as though he might have carved the Carlsbad Caverns out of the Earth single-handedly. For obvious reasons, we had nicknamed him the "Big O."

The O was one day ahead of me in flight training and had, therefore, completed his PS-1 the previous day. The O, now a "highly experienced aviator," had given me the "gouge," the inside scoop. He had provided me with that valuable information that was not written down anywhere, but essential to survival in flight school and success in naval aviation. Just the day before, the O had warned me, "If you start feeling airsick, simply unzip the side zippers to your flight suit, and then direct the aircraft air vents directly towards the openings." I now realized that opening these zippers would allow cooling air, such as it was, to more directly enter the body area of the flight suit, circulate, and cool and comfort my body. This, hopefully, would help calm my upset stomach.

Opposite page: The Rotating Service Structure (RSS) is rolled back at Launch Pad 39B to reveal in full the space shuttle *Endeavour* prior to the launch of STS-72. Rollback of the RSS marks a major milestone in the approximately three-day-long shuttle launch countdown process.

Remembering the O's advice, I tugged at my suit zippers, slid them downward and aimed the cockpit air vents towards the zipper openings.

> I was really feeling badly, and beginning to think,
> *"I'm not going to make it."*

Ah, just a slight bit better. But, I still wish we'd stop this nonsense!

Power . . . push over and negative g . . . speed, pull up and positive g . . .

We started the entire sequence all over again, but this time, on the "pull-up," the nose continued to go higher and higher, until we were upside down at the top of a full loop. At that point, with the aircraft upside down, and with me hot and sweaty, my stomach began gurgling and cramping.

The crew of Mission STS-87 depart from the Operations and Checkout Building en route to Launch Pad 39B, where the space shuttle *Columbia* awaits liftoff.

I was really feeling badly and beginning to think, *I'm not going to make it.*

Where have you gone Sky King? I'm not sure I can find you . . .

Now in the throes of a nauseating agony, I managed to double-check that my shoulder harness was locked and seat belt secured.

I braced myself . . .

As the aircraft's nose continued down towards the Earth, we again picked up speed. My stomach was tight and on the verge of eruption. Then as we leveled off, my stomach was in my throat. I just new I was going to be sick.

During this moment of unparalleled nausea, I still managed to envision the other flight students pointing and whispering, *He couldn't hack the program.* I reached down into my right flight suit leg pocket and pulled out my airsickness bag. (At least, I remembered to bring one. Without an airsickness bag, I might have had to use one of my flight gloves – not a great option, but not as bad as messing up the aircraft cockpit!) I opened the bag and slowly began to bring it up, towards my mouth. I was attempting to keep the bag in front of my chest to conceal, for as long as possible, it's presence from Peterson (whom I was now mentally referring to as "Attila") sitting comfortably in the rear cockpit. We had finally completed this, yet another loop. As if on cue, I keyed the ICS all the time attempting to sound enthusiastic.

"That was fantastic!" I lied again. "Let's do another one."

And just when I was about to lose it all, he said, "Sorry, time's up. We've got to go home."

Attempting to inject just the right amount of disappointment into my voice (and with unbridled nerve), I cried, "Aw, I really enjoyed that. I want to do more!"

I'm coming to join you Sky King.
Leave one of the bad guys for me.

Now I do not consider myself an overly religious person, but the thoughts in back of my mind would have made the staunchest evangelist proud. In my mind I yelled, *"Thank you Lord, praise God, hallelujah"!*

My stomach settled a bit on the straight and level flight home. Peterson again turned the aircraft controls over to me for a short while.

After landing, but before our post-flight debriefing, I excused myself and spent a few minutes regaining my composure in the head (restroom). I was careful to assure Peterson that I would only be gone for a moment or two, and that ours had been a truly wonderful and enjoyable flight.

During the post-flight briefing, Peterson sat down with me and gave me a thorough review of what we had accomplished during this flight, and what I should be prepared to do during tomorrow's flight. He then filled out a grade form for me. I received all average grades. Not a bad start at all. He said, "Good hop. I'll see you tomorrow," shook my hand and walked away.

I stayed for a while, looking at the grade form. On the front of the form were my numerical marks. All average. On the back of it, I noticed Peterson had written, in big script, "No apparent signs of airsickness."

I'm coming to join you Sky King. Leave one of the bad guys for me.

— Chapter 2 —

I Never Called Him Uncle Willie

"*E*NDEAVOUR, HOUSTON," CRACKLED FROM THE REAR speaker on the flight deck. This was a standard radio call to the space shuttle *Endeavour*. Mission Control at Johnson Space Center, in Houston, Texas, was trying to get our attention.

I reached for the microphone to give a response. Floating in mid-air, because of the weightlessness of space, the mic, its long cord trailing behind it, resembled a snake. The sunlight streaming in through the overhead windows caused the mic to cast an eerie, serpentine shadow. I stretched my arm higher, causing my body to pull gently against the seat belt that kept me firmly in the seat, preventing me from floating around the cabin. Just a bit more reach and I grabbed the floating reptile, depressed the spring-loaded button on its head, and responded, "Go ahead, Houston."

"Yes Winston, we show ourselves losing comm (communications) in about two minutes. We'll pick you up on the eastern satellite in just a little while."

"Roger that, Houston. We'll talk to you on the east," I replied.

I had been a veteran naval aviator and aerospace engineering officer before being selected by NASA to join its astronaut corps in the summer of 1992. And here I was, in January of 1996, on my first space flight – a rookie astronaut, thrilled to be a member of an elite team of aviators now working in space. I was assigned as Mission Specialist-2 (MS-2), the flight engineer.

I was scheduled to perform an EVA – an extravehicular activity – the technical term for a space walk. I would don an extraordinarily complex and bulky space suit called the extravehicular mobility unit, or EMU, exit the space shuttle, and work outside in the open environment of space. I was scheduled to perform a large list of tasks as we orbited the Earth at a speed approaching 18,000 miles per hour. Our crew had trained long and hard for this flight, as astronauts do for every space flight. To say the least, I was excited. I was especially looking forward to joining the group of astronauts who have been fortunate enough to have walked in space.

When I had finished my conversation with Mission Control, I released the button on the mic casing and attached the microphone to a Velcro patch on one of *Endeavour*'s rear consoles. In space, everything must be anchored in one manner or another. Otherwise, objects would float anywhere and everywhere, as evidenced by the earlier acrobatics of my slithery, microphonic friend.

The space shuttle *Endeavour* lights up the night sky as it thunders aloft from Launch Pad 39B. Liftoff of mission STS-72 occurred at 4:41 a.m. EST, January 11, the 74th shuttle mission and the 10th flight of the orbiter *Endeavour*.

Reflections From Earth Orbit

I released my lap belt and gently pushed off of *Endeavour*'s center console, very careful not to bump any of the assorted switches and controls located there. I allowed myself to glide slowly upwards, to get a look out the rear windows. We had oriented the shuttle tail-first; we were moving "backwards" through space. Now, looking out the rear windows, I could actually see what was ahead of me.

Everything within sight – the Earth, stars, Moon, and Sun – were all incredibly sharp and brilliant. The lack of atmosphere in space makes these celestial bodies appear much clearer than they do from Earth, where we have to look upward through more than 30 miles of air. Some stars actually appear to be closer to Earth than others. From orbit, space looks more three-dimensional, as it actually is, rather than the flat, two-dimensional appearance we typically get when we look at the night sky from the ground.

Beneath me now was darkness. We had moved around to the night half of our orbit. Because the space shuttle orbits the Earth once every ninety minutes, we see forty-five minutes of daylight followed by forty-five minutes of night time, over and over again. During one twenty-four-hour day on Earth, astronauts in orbit see sixteen sunrises and sunsets. It's incredible how, one moment, you're staring out of the space shuttle windows at the bright stars against a pitch-black sky, and then suddenly a brilliant beam of light peeks over the curved horizon. The sunbeam quickly blossoms into a full ball of magnificent white-hot luminescence, so bright that it forces all who see it to cover their eyes and turn their heads. A few minutes later, when the fireball has moved overhead, the mountains, oceans and deserts of the globe below are in full view. Then, equally as quickly, the darkness arrives again, yielding the now-familiar, yet unbelievable, universe of stars, moons, and planets. Below me now, in the night, I could make out thunderclouds as jagged lightning rippled inside them, illuminating them with incredible power. Over the great distance that separated me from those clouds, the awesome power of nature manifested itself in electrical form, and impressed me with this most exciting light show.

It's amazing – here, in space, I have to look *down* to see weather. It's all beneath me.

I made it, I said to myself. *I'm actually here, in space*, I thought as I gazed out of the windows. I had come a long way from those boyhood days in Coconut Grove, the all-black section of Miami, Florida.

The Grove was home and it felt good.

"The Grove," as our part of Miami was commonly known, was a thriving neighborhood, or so we thought. It was built up around Grand Avenue, the two-lane main street, and the residents lived in racially segregated conditions. The black businesses and the black neighborhood spanned several blocks on either side of Grand Avenue for about a mile. There were a couple of drug stores, a funeral parlor, barber shops, a dry cleaning shop, a movie theater, and at least one of every other business found in the larger city. Here they existed entirely in blackface. Black residents rarely needed to venture outside of this section of Coconut Grove to obtain the necessities or luxuries of life.

The Grove was comfortable and familiar to the kids who lived there. From an early age, we could walk or ride our bicycles all over the neighborhood, to and from school, to any one of the stores located there, to Virrick Park, our friend's houses, or perhaps the old Ace Movie Theater, and never lose the sense of security that comes from being in familiar surroundings. Neighbors knew each other by name. And the kids I went to school with were the same kids with whom I attended church, and played ball, and went to Boy Scout meetings. We could stop at Joe's store on the way to school in the morning and buy glazed donuts for a nickel each. On the way home from school in the afternoon we could stop at Dunn's drug store and buy soft drinks. The nickel used for a glazed donut could just as easily have been spent on a dill pickle, a Butternut candy bar, or five Mary Jane candies. The Grove had its rough element but had not yet been invaded by illegal drugs. For the most part, it felt safe and secure. The Grove was home and it felt good.

George Washington Carver School was located on Grand Avenue right at the edge of our neighborhood. Carver hosted grades kindergarten through twelve, and served only black children. Black students in grades 7 through 12 who lived in South Miami and Richmond Heights also attended compact but comfortable Carver. These students were bussed from miles away, past "white" schools, to attend all-black Carver.

I attended Carver Elementary from first through fourth grade. I then completed fifth and sixth grades at the brand new (but still racially segregated) Francis S. Tucker Elementary School when it opened to serve students closest to its location. I then returned to Carver for junior high school grades seven through nine.

Junior high school (nowadays it's called middle school) was the time in which students began to discover much about themselves. It was during the junior high school years that we began to play competitive sports, take music lessons, try out for the junior varsity cheer leading squad, and experience a host of other extracurricular activities. Some junior high students discovered a love for writing, while others joined clubs, drama, or chorus. Whatever a student's interests, junior high was the time during which he or she could explore and begin to develop those interests. Junior high was also the time when a student's activities began to contribute to the direction that an individual's life might ultimately take.

It was during my junior high school years at Carver that I discovered my talent and love for music. It didn't happen all at once, but when the jazz bug bit me – and when I discovered the popularity that musicians can command – I began to spend considerable time practicing my trumpet. I began to excel at playing the trumpet. By eighth grade I had become one of four lead trumpet players, the youngest of the four, performing regularly with the Carver Junior / Senior High School marching and concert bands. I was arranging music for the marching band by the time I entered ninth grade.

During my ninth grade year in school, the decision was made to change the historically black senior high schools in Miami to junior high school status. As a consequence, my friends and I suddenly found ourselves beginning our tenth-grade year at the giant, historically white, Coral Gables Senior High School. I vividly recall our discussions regarding this prospect. Most of us accepted the coming change with a little concern and with much excitement. We were young students, but we were old enough to wonder if we would be accepted and embraced at "the Gables," or if we would spend our days fighting those who would abuse us with racial slurs. Our parents and teachers had told us that the coming integration was a good thing. So many people had fought so hard and for so long to make integration a reality. And they told us that we were getting a better opportunity for a good education.

Upon learning that I would attend the Gables in the upcoming school year, it seemed only natural that I would try out for the Coral Gables Senior High School band.

I remember the day I rode my bicycle to the Gables for my band audition. It was a typical warm, sunny Miami afternoon during the summer before my sophomore year. Hanging my trumpet case by its handle over the handlebars of my bike, to free both hands for steering, I pedaled the mile or so from Black Coconut Grove to all-white Coral Gables. As I rode, I wondered whether I would be welcomed at the Gables with open arms or with a phony smile. It was impossible not to wonder whether my audition would be fair or a waste of time. But I was excited and optimistic, as youngsters typically are. I pedaled on into what seemed to be a new world.

Arriving at the school, I was awed by its size. Large enough to house three thousand students, Coral Gables seemed more like a small college, than a high school.

The Gables High band director was a man named William "Bill" LeDue. Mr. LeDue struck me as a cordial, but no-nonsense kind of man. He welcomed me with a handshake, but then began to question me about my previous musical experience. Then came the audition. He handed me the first-trumpet part to one of the school's fight songs, *Dixie*. This was a time before the days of the African-American peoples' expressions of the negative symbolism of *Dixie*. The

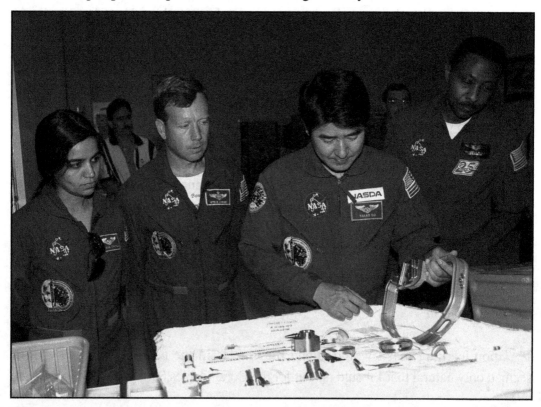

STS-87 astronaut crew members participate in the Crew Equipment Integration Test (CEIT) in Kennedy Space Center's Vertical Processing Facility. From left are Mission Specialist Kulpana Chawla, Ph.D.; Pilot Steven Lindsey; Mission Specialist Takao Doi, Ph.D., of the National Space Development Agency of Japan; and Mission Specialist Winston Scott. The CEIT gives astronauts an opportunity to get a hands-on look at the payloads with which they will be working in orbit.

arrangement of *Dixie* that Mr. LeDue gave me to play was a rousing interpretation of the song consisting of rapid eighth note and sixteenth note runs in the high octave of the trumpet register. Carver school had prepared me well. I had no difficulty sight reading the piece and with Mr. LeDue's permission, playing the "reel" section an octave lower (I had not played much during the summer and had not maintained much upper register endurance on the trumpet). Mr. LeDue, without any elaboration, simply thanked me and told me he would see me in band rehearsal at the beginning of the school year. Had I passed the audition or had I been judged not good enough to be in the band? What's going on here? I wondered. He then informed me that I should register for seventh period band. Seventh period was when the Gables top band rehearsed. He also told me in a somewhat nonchalant manner, "I need someone strong on the lower parts. If you play third trumpet this year, you'll be playing first trumpet next year." I was elated because I had performed well. I also felt a little uneasy, because now I was to enter an integrated world that held who-knows-what for me. As I pedaled my bike home, I wondered what it would be like in the Gables band. How many other students from Carver would be in the band? What style of music would we play? Would we play traditional marches or would we play the soul music I was so accustomed to playing at Carver? I thought my parents would be proud. It seemed that I was off to a good start at the Gables.

Though a relatively small school in size, my familiar Carver boasted some very good musicians, but its band was small in numbers. On a good day, we might muster forty band members. Coral Gables High School, on the other hand, was so large that there were two bands. There was the Cavalier Band, which consisted of fine musicians, but musicians who were not as proficient as other band students. Also, musicians who couldn't march, or for one reason or another found it difficult to march, could be members of the "Cav Band" portion of the school's music program. The Cav Band played in the stands at the football games, but typically did not march in the half-time shows. As I recall, the Cav Band had one hundred members or more!

Then, there was the Coral Gables High School "Band of Distinction." This band consisted of well over one hundred members. These students were among the finest high school musicians in the state, and very likely in the country. This was the award-winning band that took the field for half-time shows week after week at the school football games. This was also the group that had won nothing short of straight grades of "Superior" at district and state contests for more years than most people could remember. The musicians (and thus the band) were so good that we would only wear medals won for competition at the state level. There was not enough room on each member's band uniform for both district level achievement medals and state level medals won for solo and ensemble competition. And there was no shortage of state level medals won by these students. The medal-laden uniforms were impressive enough to make any veteran military war hero take notice.

The Coral Gables High School band was the beginning of a very profound experience for me. I can still remember the first note of my first rehearsal. At that first chord, the full, rich sound of this gigantic band was so big and powerful that it actually startled me. I wondered if anyone noticed me flinch at that first chord. I had never heard a sound like that before in my young life.

> **I remember being so unhappy that I thought of dropping out of school.**

The sound of the band was not the only thing that was new to me. The makeup of the band also required an adjustment on my part. From my first rehearsal through the next three high school years I was the only African-American member of the "Band of Distinction." Though proud to have been selected for the band, I felt as though my culture had been stripped from me. I no longer played "my" music, the jazz and rhythm and blues that I so loved. I was no longer among "my" people, my black band friends with whom I had spent the previous three years at Carver. All of the other African-American students from Carver who chose to participate in the Gables' band program were in the Cav Band. Some of these students simply were not proficient enough for the Band of Distinction. One Black student stated several times to me the he passed the audition but did not want to be a member of the Band of Distinction. He said that he enjoyed the Cav Band because it didn't require the work required in the Band of Distinction. Regardless of whether or not his assertion was true, I felt lonely and isolated. My white band mates were friendly and accepting, but I felt alone in this crowd of new peers, as if I really didn't belong. It was in this setting – the only African-American member of the band at a newly integrated school – that I settled down to play third trumpet as a "Coral Gables Cavalier."

I was certainly the most noticeable member of the band. In spite of this, I began to adjust. Tenth grade went by quickly. I did well at Gables that year, continuing to play in the band and making decent grades. I made new friends and actually enjoyed the atmosphere at school. I even won a spot on the junior varsity basketball team.

My junior year was a different story. Looking back, I don't quite understand the emotional turmoil I felt. In spite of the honor of being a first trumpet player, I can remember being frustrated and angry. I didn't enjoy the band and didn't enjoy school. I did not have that feeling of accomplishment and personal satisfaction that I had previously felt as a result of playing music. I felt anonymous and unimportant. I felt as though I wasn't accomplishing anything, not achieving anything of consequence. I didn't talk about these tumultuous feelings at home, but I remember being so unhappy that I thought of dropping out of school. Several of my peers from the Grove had already dropped out of school. I might have done so myself had I not feared for my life. I knew what my father's reaction would be if I had even suggested leaving high school!

These emotionally difficult times were certainly not prompted by the behavior of those around me. In fact, Mr. LeDue remained his same no-nonsense but cordial, supportive, and encouraging self. By now, most of the white and black students in both bands called him "Uncle Willie." There was one specific occasion, when I overheard one of my black friends in the Cav Band call Mr. LeDue "Uncle Willie." I remembered the times he awarded me demerits for rules infractions. I remembered the strict discipline with which he administered the band, and his unwavering no-nonsense demeanor. I thought angrily, *I'm not calling him Uncle Willie. That old white man is not my uncle!*

Was I going through the typical heavy emotional changes through which all teenagers pass? I don't know. I do know that I began to skip band rehearsals. I was suspended from the band more than once and assigned a grade of F in music on my report card on each occasion. Mr. LeDue talked with me and even sent me to the Principal's office on a couple of occasions for counseling. It didn't help. I was angry and unhappy. I finished that school year as a very depressed and distraught student. I don't know why Mr. LeDue didn't simply give up on me. By every right, he could have asked me not to return to the band my senior year.

Summer passed and the next fall, the beginning of my senior year in high school, I did return to the band. I returned to school and to the band with the realization that this was my last chance. Perhaps I had matured enough to realize that I had to do better than I had done during my junior year. I would graduate at the end of this current year, and if I intended to go to college and "make something of myself," I had better shape up.

At the same time that I was enjoying music, I was also aware of my strong love for technology. Devices and gadgets, electrical and mechanical, had always fascinated me. While in elementary school, I had taught myself to read electrical schematic diagrams and construct simple circuits using batteries, wires, and light bulbs. I built model airplanes out of whatever scrap materials I could find. Until sixth grade, I used to open up my toys to see how they worked. I remember cutting up our string of Christmas tree lights and using the parts in my electrical projects and circuits. I was the perfect example of a youngster who should have been exposed early on to science, engineering, and technology. There was no exposure however, for my peers and I. There was no one to introduce us to engineering or physics. I had always thought I would like to learn "something about electricity or

The STS-87 flight crew enjoys the traditional pre-liftoff breakfast in the crew quarters of the Operations and Checkout Building, following which, after a weather briefing, the flight crew was fitted with their launch and entry suits and then departed for Launch Pad 39B.

electronics" in college, but I had no idea what or how. I only knew for sure that the one thing that I did well was music. It seemed to me that if I was going to attend college, I had better major in music.

> # I had also begun to realize my own potential for leadership, accomplishment, and success.

I approached my senior year with a new attitude. I was determined to make the best of it, to do the best that I could. And this positive approach paid off. Towards the end of the first semester, I realized I was having fun. Once again I began to enjoy school and enjoy the band. Mr. LeDue not only assigned me to play the lead trumpet parts of our music, but he also designated me as section leader. As section leader I was the student coordinator, or captain, for the trumpet section. I actually led the section rehearsals, which were held to give the trumpets an opportunity to practice our musical parts alone as a section, without the remainder of the band. In addition to this honor, I earned two superior awards for ensemble performances at district competition, and one superior award for ensemble competition at the state level. Of course, the Band of Distinction won superior awards for both district and state competition.

I was busy outside of school playing gigs on the weekends with local rhythm and blues or soul bands. I became quite a popular musician among my friends, undoubtedly boosting my morale and self-confidence.

The highlight of my senior year came at a home football game during which Mr. LeDue and the Band of Distinction had to be out of town. To fulfill the band's football game obligation, Mr. LeDue placed me in charge of the Cavalier Band. The Cav Band, under my direction, was to assemble at the school, board busses, attend the football game, and perform music from the stands throughout the game, then return to the school. I was to be in charge of the musicians and the music from the time we left the school, throughout the football game, to the time we returned to the school. I was responsible for every aspect of the evening from conducting the music to issuing demerits for disciplinary infractions. This was a big responsibility and a bold statement of Mr. LeDue's confidence in me. The night was a great success for the football team, for the Cav Band, and for me. It was a great boost to my self confidence. Earlier in the school year I had earned the respect of my peers as a performer, but that night I realized that I had gained their respect as a leader. I had also begun to realize my own potential for leadership, accomplishment, and success.

As my senior year progressed, my thoughts turned to college. I had no idea which university I might attend. Mr. LeDue asked me about my plans and encouraged me to apply to Florida State University. Even then, FSU had one of the finest music schools in the country.

> # I would not be going to FSU, and that was that.

I applied to FSU. After a few weeks, I received notification that I had not been accepted. Although I was disappointed, I was not surprised and certainly not devastated. When you grew up Black in the 1960's, you came to expect this sort of reaction to your efforts. Black youths were conditioned by life to expect rejection, or at best, leftovers. And in the Black section of Coconut Grove rejection came to us all. Often. Rejections of applications for jobs, training programs, and colleges were commonplace. Of the small group of my peers who went to college, most did so as athletes. Most of those athletes attended small colleges with small prestige and even smaller budgets. I could get a fine education at any one of these smaller schools, I reasoned, so there was no need for me to worry too much about not being accepted to FSU.

In the ensuing days and weeks, Mr. LeDue occasionally asked me about FSU. I was embarrassed that I had not been accepted. I told him on more than one occasion that I had not received a response from the university. I knew that this stalling tactic would only work for so long, and that I would soon have to tell him the truth. One day while in the band rehearsal hall, Mr. LeDue once again asked me if I had heard from FSU. I finally admitted to him that I had heard from FSU and I had not been admitted.

This news did not seem to have any appreciable effect on him at all. He maintained his usual no non-sense, business-like manner, and simply said, "Come with me."

Mr. LeDue led me out of the rehearsal hall and to his office. While we walked he asked me if I still had a B average. I replied, "yes," although I actually imagined myself to be more a B-C student. I knew I really had not worked up to my true potential. I didn't know what to expect. I figured Mr. LeDue was disappointed in me, and the walk to his office would give him the opportunity to tell me so. Having arrived at this conclusion, I was at least relieved that he would lecture me in private, so that other students in the band hall would not overhear. I just wanted it all to be over. I could certainly go to school some place, even if the school I ultimately attended did not have the prestige of Florida State University.

Mr. LeDue's office consisted of a typical workspace complete with desk, chair, telephone and outer foyer. The door that separated the foyer from the office space was constructed with a large pane of glass in its upper half. Mr. LeDue told me to wait in the foyer while he went into the inner office and closed the door. I could see through the glass in the door that separated the two areas, but I could not hear what was being said inside.

I watched Mr. LeDue pick up the telephone and dial. I could see that he was talking with someone, for his lips moved. During the exchange, he nodded his head. Meanwhile, I just stood there in the foyer quietly, waiting and wondering what Mr. LeDue was doing. Becoming a little nervous and uncomfortable, I paced the small foyer hoping that whatever was happening soon be over. Who is he talking to? I wondered. Is he speaking with the main office? Is he calling to check on my grades? I would not be going to FSU, and that was that. There was nothing anyone could do about it. Mr. LeDue finished his conversation, put down the telephone receiver, and emerged from his office. He simply told me to go to my next class.

Two or three days later I received a letter that said, "You are hereby accepted to Florida State University." It was absolutely incredible. I was shocked, excited, elated, a little scared, but ready for this next challenge. When I told Mr. LeDue the good news he remained his same cordial, no-nonsense self but this time, he smiled, and said, "congratulations."

I never missed an opportunity to let my relatives, friends, and neighbors know that I was going to attend Florida State University. For a little black kid from the disadvantaged section of Coconut Grove, acceptance to a major university like FSU was a big deal! It was especially rare for someone from my neighborhood to attend any university without doing so on an athletic scholarship.

Florida State University in the 1960's presented its own set of challenges. The recently integrated campus was not a particularly accepting and nurturing one for Black students. FSU did prove to be exciting, academically challenging and broadening for me. The university introduced me to many new ideas, opportunities and friendships. I excelled in my music studies. I was the only freshman in the university jazz orchestra. I was a student composer and arranger for the jazz orchestra and also for several of my own bands that I put together over my four years at FSU. I planned, produced and hosted my own weeknight radio program on WFSU-FM, the National Public Radio affiliate located on campus. In addition to studying music, I discovered engineering. In doing so, my lifelong love of technology was rekindled. I took classes in calculus, physics and electronics in addition to my music classes.

FSU caused and permitted me to learn, grow and mature to the point where I realized that I could accomplish anything that I set out to achieve. I recognized the limitless potential existing within me, and became acquainted with the limitless possibilities existing before me. All I had to do was decide what I wanted to do in life – and then do it.

Astronaut Winston E. Scott performing EVA activities during the STS-72 mission, his first space flight.

I graduated from FSU and spent the next twenty-seven years of my life as a Naval Aviator, engineer and astronaut.

And he didn't have to make that call.

I often reflect on the story of Mr. LeDue's phone call and its effects on the direction my life ultimately took. I think about how no one achieves any measure of success totally on his or her own, but instead with the help of other people. I think of how we are sometimes helped by the people we least expect, and that the extent of that help is sometimes far greater than we would ever have expected. I also think about how much the true benefit of an education is not in what's learned from textbooks, but rather in what one learns about oneself. And it's what we know about ourselves, inside, that gives us the courage to be all that we can be.

I think about the three years I spent at Coral Gables High School. I think about the Band of Distinction, the football games, and the concerts. I think of Mr. LeDue and the interest he took in me. I think of how he never lost confidence in me, never gave up trying to help me, and never punished me for poor conduct. Oh, he certainly issued consequences! But he never punished me. I remember how the other kids called him "Uncle Willie." But I never called him Uncle Willie. He was always "Mr. LeDue" to me. After all these years, I guess he'll always be "Mr. LeDue."

I am convinced that my life was at a crossroads on that day, when Mr. LeDue took me into his office to make a special phone call. I am convinced that the course of my life was changed when he spoke to the university official on the other end of the phone line on my behalf. Had Mr. LeDue not made that telephone call, I would most likely not have entered the navy, become an aviator, engineer, and astronaut.

And he didn't have to make that call. It would have been much easier, and just as correct, for him to have simply let things stand as they were.

So, here's to you, Mr. LeDue. Thank you for believing in me even when I sometimes was not the most cooperative, or grateful, student in your band. Thank you for caring about me to an extent well beyond the scope of your teaching duties. Thank you for taking an interest in me, for challenging me, and for demonstrating the confidence that you had in me. Most of all, thank you for taking that extra step which changed the entire course of my life.

Here's to you, Mr. LeDue.

No, . . . perhaps this one time – here's to you Uncle Willie.

Thanks.

STS-87 Mission Specialist Winston Scott is assisted with his ascent and re-entry flight suit in the white room at Launch Pad 39B by Danny Wyatt, NASA quality assurance specialist.

— Chapter 3 —

And What May Come of Dreams – Ascent

IT SORT OF "LEAPS" OFF THE PAD AT LIFT-OFF, NOT AT ALL LIKE the slow motion pictures you see in the movies and on television. When the space shuttle blasts off from the launch pad it kicks you in the backside, and by the time you pass the top of the launch tower you're already traveling faster than one hundred miles an hour.

I remember waiting for my turn to climb onto *Endeavour*'s flight deck before my first flight into space. We had arrived at the launch pad some three hours prior to scheduled lift-off. Around midnight, in January, the weather, even in Florida, was cold. In fact, this particular night had the coldest weather for a space shuttle launch since the day we lost *Challenger* and her crew. During our preflight news interviews, the media seemingly spared no effort in making us aware of that fact. The frigid temperatures were accentuated by the fact that the shuttle was standing upright on its tail, and the hatch leading to the crew compartment was one hundred and ninety-five feet in the air. Despite the cold, we were warm; our bulky orange launch and entry suits (LES's) insulated us well.

One by one we climbed into the crew compartment trying not to step on any delicate flight equipment, kick any switches, or damage any displays. Protective panels and switch guards cover much of the cockpit equipment before launch, but we exercised great care so as not to inflict any damage on the machine that was soon to transport us to low-Earth orbit and back.

The Astronaut Support Person (ASP) assigned to our flight assisted us as we donned our parachutes and survival vests, strapped ourselves into our seats, and put on our helmets and gloves. We had to have help climbing into the orbiter. Its

vertical orientation caused us to have to literally *climb* into our seats and lie on our backs with our feet up in the air. You actually lie back looking up and out of the front windscreen towards the sky. We climbed in and tried to settle down but there was no getting comfortable. Once strapped in, we spent the remaining couple of hours before lift-off reviewing our procedures and conducting systems checks with the Launch Control Center (LCC). We checked every system imaginable – suit oxygen, suit pressurization, orbiter pressurization, orbiter hatch seal integrity, orbiter communications, shuttle computers . . . You name it, we tested it.

The hours just prior to lift-off can be busy with crew activities; or they can be quiet for the crew while the LCC conducts tests. It's a time during which an astronaut can sit quietly and reflect on what he or she is about to do. Or it is a time when an astronaut can just lie back and have a snack if hungry. I remember lying on my back hoping that we wouldn't have an abort. I didn't want to have to come back tomorrow night and go through the whole ordeal of suiting up, climbing up, strapping in, and lying here again, hoping to launch. I thought of my family members and friends who had traveled from all over the country for this event. I didn't want them to have to either go home without seeing me launch or make the decision to remain in town another day and hope for better luck the next night. I also remember feeling a bit sleepy. The many long days of training and the anticipation for what was about to become the most incredible trip of my life had left me somewhat sleep-deprived. I thought, *If this wasn't so serious, I could take a nap.* I wasn't serious about taking a nap, because I was psyched up about this mission. I was pumped up and ready! I thought, *Let's wake this monster up and get going.*

The countdown clock passed T-5:00 – five minutes prior to lift-off. I followed the checklist as Brent, our pilot, started the auxiliary power units and checked the hydraulic pressure. *This will nudge the sleeping beast*, I thought. Moving my head to the left just a bit was awkward in the LES, but doing so allowed me to see the hydraulic pressure gauges too. *Three green!* I noted. At T-3:25, three minutes and twenty-five seconds before lift-off, I felt the motion of the main engines automatically swinging through their test positions then settling into launch position. *The beast is stirring now. He'll awaken soon and we had all better be ready!*

Opposite page – The crew of STS-72: (a) Brian Duffy, Commander; (b) Brent W. Jett, Pilot; (c) Leroy Chiao, Mission Specialist; (d) Daniel T. Barry, Mission Specialist; (e) Winston E. Scott, Mission Specialist; (f) Koichi Wakata, Mission Specialist.

At T-2:30, the LCC instructed, "*Endeavour*, clear Caution and Warning panel and configure computer displays."

Brent complied and responded, "Caution and warning panel cleared and displays configured."

> Adrenaline pumping, I whooped! as we leapt off the pad.

At T-2:00, the LCC announced, "*Endeavour*, start O₂ flow and close and lock visors." We followed instructions and were soon fully enclosed inside our LES's. I could feel the orbiter cooling system circulating chilled water throughout my LES. It may have been cold outside, but inside *Endeavour*, in our LES's, we would be dangerously hot if not for the orbiter systems keeping us cool. The cool flow to my body was accompanied by the sanitary smell of shuttle-manufactured air. I watched the clock count down: ten, nine, eight, seven seconds; here the ground launch sequencer computer was in control of the launch sequence. I realized that, unlike the jet fighters that I was so accustomed to piloting and controlling, here inside the space shuttle the crew was just along for the ride. I saw the space shuttle main engine (SSME) gauges come alive and heard the engines ignite. *The beast is awake now!* There was a brief sputter then a smooth increase to maximum power. Spectators outside would be marveling at the deafening sound, but inside the shuttle, with helmet on and visor closed, my LES efficiently muffled the engines' roar so that I could clearly hear the shuttle radios and intercom. The orbiter swayed back and forth, straining against the hold-down bolts that locked it to the launch pad. *He'll break his leash if we don't release him soon!* With all three SSME's working properly, the solid rocket boosters exploded to life and sent smoke and bright flame up and around the forward windscreen. It was as though we were sitting inside a big fireball and, for a moment, a brief moment, I thought, *What have I gotten myself into?* "Three . . . two . . . one . . ." Explosive charges fired severing the hold-down nuts and releasing the demon. It screamed off the pad, kicking, bucking, and clawing for altitude. Adrenaline pumping, I whooped! as we leapt off the pad, finally doing what we had been training to do for so long – rocketing into outer space. What a ride!

Opposite page – This distinctive view of the space shuttle *Endeavour* was taken from the Rotating Service Structure (RSS) as it was being rolled away from the vehicle at Launch Pad 39B prior to the launch of STS-72.

Reflections From Earth Orbit

> **Those moments that can ultimately shape the direction of a person's life may last only a few seconds.**

It's not a smooth ride. During the first two minutes of flight, the first stage – seven and a half million pounds of thrust propelling the shuttle skyward – produces a bumpy, rocky ride. Shaking and vibrating, the noise, smoke, flame and acceleration of the shuttle monster leaves little doubt in your mind that you are going somewhere fast. But you are confident, at least somewhat, that where you're going is where you want to go. As flight engineer, I was to assist the commander and pilot in monitoring shuttle systems using the computer screens, systems dials, and performance gauges. As we rumbled upward, faster and faster, shaking and vibrating, I thought, *I'm supposed to read these gauges?* Past the launch tower now, the computers rolled the shuttle to steer us in the proper direction of flight. This rolling / turning maneuver left us upside down. I barely noticed the upside-down attitude because our forward acceleration pushed me back into my seat. Brian, our commander, keyed his microphone and transmitted the standard short and efficient "Roll program Houston." The reply of "Roger, roll, *Endeavour*" from mission control was equally concise. Moving faster and faster, we passed mach one, the speed of sound, after only forty-three seconds.

There are thousands of steps that must be executed during a space shuttle ascent into space. The ship's computers execute them all, unless there is a malfunction. Then the crew must take over and manually fly the orbiter. In the meantime, we monitor the computers and the multitude of other displays, and remain prepared to take control if necessary. With so much going on, communications with the ground must be kept short and to the point.

After only a few seconds of flight, the shuttle's computers reduce the power from the main engines. The solid rocket boosters likewise reduce their power output at this time. This power reduction, called the "thrust bucket," is necessary in order to keep the shuttle from going too fast while in the lower, "thick" part of the atmosphere, which would exceed its structural strength limits. Exceeding structural limits could lead to destruction of the shuttle and loss of the crew. The reduction of power and resulting loss of acceleration might be disconcerting were it not for our training. We were expecting it exactly as it happened. The thrust bucket's momentary lull in acceleration makes it almost feel as if the shuttle were pausing before shifting gears, so that it could then climb higher and faster. It was as if the

Reflections From Earth Orbit

STS-72 Mission Specialist Winston E. Scott dons his launch / entry suit in the Operations and Checkout Building with assistance from a suit technician.

giant speed demon, alive and running, paused for just a moment to get its second wind, so that it could go even faster. This lull lasts only a few seconds but is very important to achieving a safe ascent. Some of the more memorable moments in our lives, those moments that can ultimately shape the direction of a person's life, may likewise last only a few seconds. Our dreams, the experts tell us, no matter how memorable and important to us, similarly last for only a few seconds.

If we were poor, we didn't know it.

I must have been ten or eleven years old when I had the dream. Our house, modest by today's standards, was modern and comfortable. Its three bedrooms and two bathrooms had been a source of excitement and pride for our family when we moved into it. At the time of the dream my sister, LaVerne, had already left for college. A freshman at sixteen years of age, she was a source of pride for our family. My younger brother, Junior, and I enjoyed a safe home with both parents present and working hard to give us a "proper upbringing." Daddy worked two jobs in order to send my sister to school. Mama worked also. We definitely were not rich; if we were poor, we didn't know it. We had all of what we needed and much of what we wanted. There was security and there was discipline, and most of all, love. It was at this point in my life that the dream occurred. Although not particularly spectacular, it must have been meaningful to me because it's still vivid in my mind some four decades later.

In this dream, I was standing in our yard at the left rear corner of the house. It was a sunny day, typical of south Florida except that everything was unrealistically bright and clear. The light green house with its white tiled roof was more spectacular than in real life. The lush dark green grass, the full leaves on the trees, and the blue sky were all unrealistically bright. I wore blue jeans that were a brighter blue and sneakers that were a whiter white than in reality. And I remember, somewhere in my dream, the color red, a weird bright red. But I don't remember whether it was my shirt, the flowers, or something else; but something was colored red. The entire dream scene was in the awesome brightness and clarity that I would later see while looking at Earth through the vacuum of space.

Opposite page – The STS-72 astronauts stand outside the space shuttle *Endeavour* at Launch Pad 39B. The flight crew is at KSC participating in the Terminal Countdown Demonstration Test, a dress rehearsal for launch.

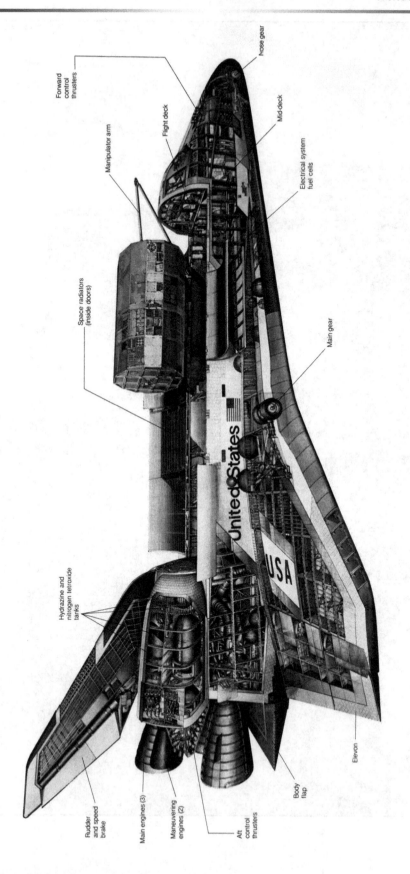

Reflections From Earth Orbit

On *Endeavour*'s flight deck, Brian transmitted, "Throttle up Houston" over the radios. "Roger *Endeavour*. Go with throttle up," was the reply. I watched the gauges rise as we came out of the thrust bucket and I felt the increased power from the engines. The acceleration pinned me back into my seat as the g-forces grew. My body weight increased, building towards three times its normal amount. The mission elapsed time (MET) clock had been counting up since lift-off. *Two minutes into flight,"* I thought; *there it is – P_C (engine chamber pressures) less than fifty percent.* I heard an explosion; *good SRB sep,* I thought as the solid rocket boosters (SRB's) were exploded away from the orbiter. The SRB recovery ships would pick up the boosters from the ocean for refurbishment and reuse. Meanwhile, we continued into space on the main engines. As the g-forces built, I moved my gloved hand to turn a page in my checklist. I expected it, had felt it before, but was still surprised at how heavy my arm felt under three g's. I raised my hand to rotate my helmet upward and tilted my head to see the evaporator output temperature gauge. My head, too, felt three times its normal weight. Faster and faster, higher and higher we rocketed towards our target – outer space.

At three minutes into the flight I announced over the intercom, "Evap Out Temp coming down." This informed the crew that the shuttle cooling system was functioning properly. "Thanks," Brian said. The roar of the main engines was steady in my headset as I moved my eyes away from my instruments. I then moved my heavy head and glanced out the overhead windows. I could see the darkness below us. I turned my head forward again and I could see darkness through the front windscreen also. And there, in the distance, was the terminator, the line separating the dark side of the Earth from the sunlit side. We would be passing into daylight soon.

Opposite page – This illustration is a shuttle orbiter cutaway view with call-outs. About the same size and weight as a DC-9 aircraft, the orbiter contains the pressurized crew compartment (which can normally carry up to seven crew members), the huge cargo bay, and the three main engines mounted on its aft end. There are three levels to the crew cabin. Uppermost is the flight deck where the commander and the pilot control the mission. The mid-deck is where the gallery, toilet, sleep stations, and storage and experiment lockers are found for the basic needs of weightless daily living. Also located in the mid-deck is the airlock hatch into the cargo bay and space beyond. It is through this hatch and airlock that astronauts go to don their space suits and manned maneuvering units in preparation for extravehicular activities, more popularly known as spacewalks. The space shuttle's cargo bay is large enough to accommodate a tour bus (60 x 15 feet or 18.3 x 4.6 meters). The cargo bay carries satellites, spacecraft, and scientific laboratories to and from Earth orbit. It is also a work station for astronauts to repair satellites, a foundation from which to erect space structures, and a hold for retrieved satellites to be returned to Earth.

In my dream, while standing in the yard, I slowly turned my head upward and looked towards the roof of the house. Then gradually and smoothly I rose into the air and began to fly. I distinctly remember not being afraid, nor was I particularly excited. I just flew. I felt comfort and satisfaction in the fact that I was airborne, under my own power, of my own will. I rose to a height about even with the roof line and began to fly from the rear of the house towards the front. I was not flying particularly fast, but when I attempted to turn in a different direction, I couldn't precisely control myself. I tried to turn left, but the turn was awkward and not crisp and exact. I tried to go higher than the roof line but couldn't. I remember only the slightest frustration at not being able to fly exactly where and how I wanted.

Inside *Endeavour* the ship's computers were controlling our ascent. We watched the velocity tape march upward. With the SRB's gone, the ride was smooth, like riding in a new car, except that we just kept going faster and faster and faster. Our altitude rose at an amazing rate; then, at just past fifty miles, three hundred thousand feet high, Brain transmitted, "Houston, we just made four new astronauts." He was referring to us, the four first-time flyers in our crew who had just passed the official threshold of space. Pinned back into my seat, my helmeted face must have shown a smile. *Now we're real astronauts*, I thought; *no matter what happens from here on, I have flown in space!*

As I flew along the roof line of our house, I felt more than thought that, in spite of a little awkwardness, I was higher than anyone else around me. *I'm doing what no one around me can do. I'm going where no one around me will ever go!*

Inside *Endeavour*'s cockpit, over the roar of the engines, we heard the reply from Houston, "Roger that *Endeavour*, and congratulations to the new astronauts."

At these extreme altitudes there is very little atmosphere and therefore the speed of sound is essentially zero (no sound gets transmitted in a vacuum). Mach number doesn't have any meaning under these conditions so, up here, we begin to measure our speed in feet per second – *thousands* of feet per second. I saw fifteen thousand, then a moment later eighteen thousand, then twenty thousand feet per second. Finally, after eight and a half minutes of flight, I saw twenty-five thousand, eight hundred feet per second. The computers throttled the main engines down reducing their output power to zero. Then came MECO – Main Engine Cut-Off. The rapid loss of thrust and acceleration threw me forward. My shoulder harness held me, restraining me in my seat.

As I flew onward, towards the front of the house, I tried again to turn and to go higher, but I couldn't. Then, as dreams often do, it suddenly ended and I woke up.

After MECO, everything around me – my checklists, helmet bag, everything – began to float. My body rose a little bit, but was restrained by my lap belt and shoulder harness. I needed to get out of my seat and position myself at the overhead windows to photograph the external fuel tank as it tumbled back into the atmosphere. Still strapped into my seat, I removed my helmet and it began to float away. I caught the helmet, removed my gloves and placed them inside the helmet. They didn't remain there and then my helmet was floating in one direction and my gloves in another. I released my lap belt to go after my helmet and gloves and immediately began to float. I felt like a clown; new to weightlessness, I was flailing about, flight gear floating all over the place. Brian, the commander and a veteran space flyer, coolly and efficiently turned around in his seat and looked straight at me. I thought, *Oh boy. I've been in space for less than a minute and already my commander is going to chew me out.* But Brian broke into a big smile and said, "Welcome to zero g." I laughed, relaxed, and retrieved my helmet and gloves and fastened them to my seat. I then caught the camera that Leroy had floated up to me from the lower (mid) deck, and took my pictures. I hadn't yet learned to precisely control my body in the weightlessness of space, but I was higher than most people had ever been. I had gone where few people had ever gone.

"We see MECO Houston," Brian transmitted. "Roger *Endeavour*, good MECO. No OMS one required."

Much has been written about dreams. Some people believe that dreams can provide warnings, explain mysteries, and even predict the future. I believe that no one – no mystic, soothsayer, palm reader or clairvoyant – can interpret a person's dreams better than the dreamer himself. I suspect most people ignore their dreams, as I ignore most of mine. But this particular dream, having remained with me all my life, couldn't be ignored.

While growing up, I was a fairly ordinary kid. I did not particularly stand out amongst my peers. And though no one would have guessed it at the time, I always felt that I had the potential to achieve something unusual, to rise to unusual heights, to do what very few people could do. Perhaps my dream foretold my future – a future of unusual achievements in unusual places.

Who can truly know the nature of dreams? Who can say what truths dreams may convey or mask? And who can possibly know precisely what may come of dreams?

The Rotating Service Structure (RSS) is rolled back at Launch Pad 39B to reveal in full the space shuttle *Endeavour*. Rollback of the RSS marks a major milestone in the approximately three-day-long shuttle launch countdown process.

— Chapter 4 —

Day Pass

ONE OF THE MOST FREQUENT QUESTIONS THAT I'M ASKED IS, "how does it look in space?" By "it," I assume the questioner means the Earth, the stars, the Moon, and everything else that can be seen from orbit. I always hesitate before answering because I've found it impossible to fully describe how the universe looks from space. Adjectives such as beautiful, magnificent, and surreal come to mind. But even as I speak them, I know that I've done my audience a disservice by not being able to give an adequate description of how "it" looks in outer space. I try however, to do just that.

The view from orbit is unlike any view that most people have ever seen. The Earth is visible as a sphere, not unlike the big globe that stood in the corner of your elementary school classroom. Unlike that plastic model, however, this globe is alive. It is enormous, floating quietly in the blackness of space, and it's breathing and revolving. This globe is surrounded by what appears to be a paper-thin atmosphere. The atmosphere, in fact, looks so thin that you could almost get the impression that it might at any moment dissipate, dissolve into nothingness, and float away into space, depriving people on the ground of their precious oxygen. This doesn't happen, of course, but the shallow depth of this life-sustaining atmosphere gives an impression of disturbing fragility.

I can vividly remember the first time I actually paid close attention to the Earth's atmosphere from space. It was also the first occasion upon which I took the time to fully absorb an orbital sunrise.

Our space shuttle *Endeavour* crew had been in orbit for a day or two – that is, one or two twenty-four-hour time periods, as opposed to one or two sunrises – when we finally took the opportunity to pause from our work and simply enjoy the view.

The STS-72 astronauts depart the Operations and Checkout Building and head for Launch Pad 39B and the first shuttle mission of 1996.

Reflections From Earth Orbit

I had been performing experiments on *Endeavour's* mid-deck with two crew mates. I had been working with the checklist booklet that served as my guide to completing the experiments. I attached the booklet to a mid-deck locker with Velcro to keep it from floating about and becoming lost somewhere in the cabin and slid my feet from under the floor-mounted foot restraints. Pushing myself along with my hands, I slowly glided through the air toward the ladder that led from the lower mid-deck to the upper flight deck. I certainly didn't need a ladder here in space. In fact, the ladder wouldn't be needed until after landing at the end of the mission, when we would climb down from *Endeavour's* crew access hatch to the ground. Emerging on the aft flight deck, I was joined by my two crew mates and we three rookie astronauts took up station at the two rear windows, gazing out at and remarking on the view.

Endeavour had been established in a tail-first, wings-level attitude. Tail-first means that we were actually flying backwards through space. Because we were flying backwards, we could see from the rear windows what was actually out in front of us, what was approaching as we moved around the Earth. With our wings level, *Endeavour's* belly pointing towards the Earth, we could see the Earth's surface out to the left and to the right, as well as in front of us. Not too far out in front of us was the Earth's curved horizon.

Most people understand that when it's daytime in one location (say in the United States), it is nighttime on the opposite side of the world (such as China), and vise versa. At speeds approaching eighteen thousand miles per hour, space ships in orbit travel around the Earth so quickly that astronauts move from daylight (over the United States) to nighttime (over China) and back to daylight again (over the United States, completing one orbit) in about ninety minutes. So, during each orbit of the Earth, astronauts experience forty-five minutes of daylight – a day pass – followed by forty-five minutes of nighttime – a night pass. From orbit, you can witness sixteen sunrises and sixteen sunsets every twenty-four hours.

Absorbed by the clarity and brilliance of the stars as we neared the end of a night pass, we hadn't given much thought to the fact that orbital sunrise would occur soon.

The communications radios were silent and so the only sounds were those of *Endeavour's* life-sustaining equipment.

55

It was then that we saw a most marvelous sight. A tiny edge of the Sun peeked over the horizon and shone through the thin atmosphere of Earth. As it did, the atmosphere acted as a prism causing the sunlight to break into a rainbow of colors. Each color was sharp and clear, and we could have counted them had we so desired. I vividly remember seeing red, orange, yellow, green and blue, spread out from top to bottom of the layer of air. Each bright, colored band of this solar-powered rainbow gleamed across space. We were speechless! Then, before we could catch our breaths, the full disc of the Sun appeared over Earth's horizon. Once it did, what had a moment before been a powerful array of color was now an indescribably bright ball of flame and fury. As the Sun's beams of solar energy attacked our ship, each one of us, simultaneously and without saying a word to one another, turned our heads and shielded our eyes. The flight deck of the space shuttle was flooded with light and we could only shake our heads in awe.

This phenomenon lasted only a moment, the Sun continuing to rise higher and higher in the orbital sky. As it did, the surface of the Earth became brighter and clearer, revealing more of its features as they appear only when visible from space.

I recall the deep blue color of the Atlantic Ocean. I could actually see the white wakes formed on the blue ocean surface behind giant ships. The ships themselves were too far away and therefore too small to be seen with the naked eye. As the vessels moved, their wakes spread for miles and became long enough and wide enough to be seen in the gleaming sunlight. These wakes resembled bright white inch-long streaks on the ocean surface. The ocean water varied in color from a dark blue at the deeper locations, to a lighter blue and sometimes even green in the shallower regions.

> ## We saw smoke from the oil fires in Kuwait.

Looking further ahead of *Endeavour,* I saw the Canary Islands, pointing our way toward Africa. In a matter of minutes we arrived over Africa. I could see the myriad beautiful colors of the Sahara Desert. I had always thought that all deserts were a single drab, light brown, much like typical beach sand. In reality, the Saharan sand exhibits an amazing array of colors ranging from an orangish-red to black. The bright reddish color, called desert varnish, is formed by the oxidation of rock. The tops of the mountains, uncovered by the blowing sand, are black and decorated with curved lines of gray soil. Wind-sculpted sand dunes are clearly visible and are so numerous that they can give the red-orange desert background the appearance of a giant orange peel. The sharp-edged sand dunes, actually miles apart, appear from space to be separated by only inches. As we flew high above the Sahara, I allowed myself a single unpleasant thought as to what it would be like to be on foot on that thoroughly isolated terrain. Though beautiful, the Sahara Desert must be deadly. As we moved further into Africa we strained to see Lake Chad, a favorite photographic subject of astronauts. "There it is!" one of us said. We could see the outline of darker, damp soil surrounding the existing pool of water, indicating how large the lake had been many years ago. We could also see from the size of the pool of water, how much smaller it has become after decades of drought. To the south of Lake Chad, on the right side of *Endeavour*'s flight path, we could see the boundary between the dry desert and the greenery of tropical Africa. A puff of white smoke became visible over a stretch of lush, dark-green jungle. We surmised this smoke might be from the controlled, or uncontrolled, burn of hundreds or even thousands of acres of forest. The thick, dark jungle reminded me of the scenery on those Saturday afternoon movies about ape-men and jungle explorers that I watched with my friends when we were children. I could now see the eastern coast of Africa approaching. In a matter of minutes, we had crossed the entire African continent and were moving out over the Indian Ocean. In the distance, off to the left of our course, was China, covered with clouds that denied visibility. This is often the case over China, and it's not very often that an astronaut gets a naked-eye view of the Chinese countryside from space.

Now, off to my right, again in the distance, I could see Australia. The entire continent was visible through a single space shuttle window. I thought, *Australia must be lonely. It's not physically connected to any other land mass.* Like Africa, the "land down under" sported light and dark colors, its beauty enticing the orbital viewer to swoop down for a visit. In just another moment we had completely passed Australia, moving rapidly around the world at a pace that would water the eyes of the most daring race car driver, and even the fastest jet pilot!

With each orbit the space shuttle traced a slightly different path over the Earth and we saw the details of a different part of our world. On other day passes we saw smoke from the oil fires in Kuwait, the snow on top of Mount Everest, and the dark-colored soil atop the Himalayas. We also saw the California coast, the Mississippi river, and entire east coast of America (from Puerto Rico through one window to Boston through another).

I looked out ahead of us again and realized that no textbook or flight simulation could have prepared me for what I saw next. I saw nighttime approaching! Still in bright sunlight, I could see the darkness of night approaching very rapidly, as if someone were pulling a gigantic blanket over the Earth in front of me. Experienced from the ground, approaching nightfall appears to us as a gradual turning down of the Sun's brightness, until its light is completely gone and we're in darkness. In orbit, we moved from sunlight, through the thin gray area of the terminator (the line between day and night), and into darkness in a matter of seconds. This dramatic change would be remembered forever by anyone fortunate enough to witness it.

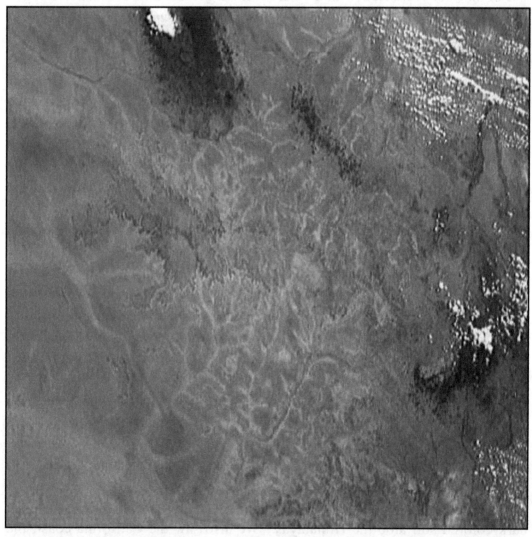

This view of the Ennedi Plateau (Chad / Sudan Border) in Africa shows the effects of severe wind erosion on this desert landscape in the far eastern portion of the Sahara Desert. The dark areas are lava flows, the calderas have long ago been eroded by the desert winds. The red toned surrounding surfaces are thick tropical soils, remnants of much earlier times when this region was a much greener savanna landscape that have also been severely eroded.

Time is precious on orbit. Our daily schedule is extremely full, and window gazing is a luxury for which we can afford only a little time. It was time to get back to work.

I pushed myself away from the window and floated toward the inter-deck ladder. Flying slowly downward, I went head-first through the inter-deck access opening, and back onto the mid-deck. I slipped my feet once again in the foot-loop restraints to keep my body stationary – I wouldn't get any work done if my body was constantly floating up and away from my experiment.

I pulled on the checklist that I had stowed earlier and heard the familiar tearing sound of the Velcro giving way. The book came loose into my hand and I opened it to the page where I'd left off. I was ready to resume my experiment, but before doing so I paused, remembered the beauty of the day pass that I had just witnessed, and uttered the veritable understatement, "This is so cool!"

The space shuttle *Endeavour* and its crew of six glide in to Runway 15 at KSC's Shuttle Landing Facility after spending nine days in space on the STS-72 mission. It was the eighth night landing of the shuttle since the program began in 1981, and the third night landing at KSC.

— *Chapter 5* —

A Model Plane For Winston

"**O**K, THAT'S ALL I HAVE. YOU GOT ANYTHING FOR ME?" I asked my RIO (Radar Intercept Officer) Lieutenant "Pink" Floyd. We were finishing the briefing and preparing for a night IR (instrument training route) flight in a Navy F-14 Tomcat fighter jet. Our plan was to take off after sunset, navigate our way at high speed and low altitude to a simulated target, conduct a mock attack on that target, and then navigate our way back again at high speed and low altitude to "home plate" for a night landing. The preflight briefing – the event during which we discussed everything that we planned to do during the upcoming flight – was, of necessity, meticulous and detailed, because mission success and crew safety depend upon every aspect of the flight being executed with precision.

"Nothing Slick. I think you covered everything. Should be a good one. Let's go see if we can't find our way to our target and back before it gets too late."

"OK," I said. "Maintenance says we have aircraft number two-thirteen, a TARPS (Tactical Air Reconnaissance Pod System) bird."

"Good," Pink replied, and then added, "I'll see you in maintenance control. I need to make a head call (bathroom stop)."

"See you in maintenance," I called back over my shoulder as I left the ready room.

I was a Navy Lieutenant Commander. I'd been flying the Tomcat for a couple of years and been in the squadron for several months. It was everything a pilot could dream of – fast flying in the U.S. Navy's premier jet fighter. This was the stuff other people paid money at a theater to watch. Here, I got paid to do it. What a life!

The F-14 Tomcat is a supersonic twin-engine, two-seat fighter plane with variable sweep wings, designed to operate at night and in all weather conditions. The F-14 can simultaneously track up to 24 targets and mounts up to six Phoenix AIM-54A missiles. Other air-intercept missiles, rockets and bombs can also be mounted.

    I left the ready room, turned to my right, and headed towards the stairwell at the end of the passageway, intending to take the stairwell from the upper level where the ready room was located down to the lower level where the aircraft were kept. As I walked along the narrow passageway, I casually looked out each large viewing window as I passed it. Through these windows I could see down onto the hangar floor where the aircraft awaiting maintenance were parked. I passed these windows numerous times on any given day and had probably glanced outward on each occasion. This time, for no particular reason, I paused and looked out the last window at the jet fighters parked below. The overhead lights inside the hangar were on. The gleam reflected from the surface of the sleek aircraft and gave them a shiny, new, and polished look. At the rear of the hangar was aircraft number 201 with an avionics panel open. Evidently the electronics technicians were working on one of the aircraft's electronic systems. In the opposite corner of the hanger I could see aircraft 214. It appeared to be in bad shape, with its upper wing panel skin peeled away. I surmised that an inspection of the attaching bolts of the wing-sweep mechanism was in progress. Then I glanced at the aircraft nearest to the window where I was standing. The lights made its skin sparkle somewhat. I admired the yellow and black paint scheme with the deadly looking skull and crossed bones painted on the vertical tails. On the side of this fighting machine was painted number 205. But the sight that made me swell with pride was what I saw painted on the canopy bow. There for all to see, in bright letters was:

LCDR. Winston E. Scott

Slick

I stared for a while and recalled the war movies I had seen as a boy growing up in Coconut Grove. In those movies the American fighter aces had each been assigned their own individual aircraft, with their names painted on them. During aerial combat these pilots would twist and turn, climb and bank, and fight their way to victory over any enemy that threatened freedom and the American way of life. On occasion, these movie warriors would even pit their aircraft against giant monsters and space aliens. But always, the movie cameras allowed the audience a view of the cockpit and the hero inside, with his name and rank prominently displayed.

> ## African-American residents of the Grove traditionally didn't venture east of MacDonald Street.

As I stood there in the squadron passageway, remembering such childish things, I thought the shiny Tomcats below resembled the sleek plastic model airplanes that I'd built as a boy. Constructing model airplanes was not a favorite pastime for little black boys in segregated Coconut Grove, but my brother Junior and I certainly built our share. Our favorite store in the Grove was the five- and ten-cent store. There wasn't much there at that time that could actually be purchased for five or ten cents, but there were wondrous treasures that could be had for a relatively small amount of money. Junior and I especially liked the small plastic model airplane kits. For ninety-eight cents (plus three cents tax) we could buy a realistic model kit of an F-86 Saber jet, a P-51 Mustang, a Navy Banshee, or any of a variety of other military fighter aircraft. For an additional ten cents we could buy a tube of plastic cement with which to assemble the model airplane once we got it home. Junior and I spent many hours assembling and "flying" our fleet of military aircraft, a fleet of which any navy admiral or air force general would have been proud. In contrast, however, I cannot remember a single instance in which any of our neighborhood buddies purchased a model airplane kit. Our neighborhood friends were fascinated by airplanes, as all little boys were, but only Junior and I bought and assembled model airplanes.

African-American residents of the Grove traditionally didn't venture east of MacDonald Street. The park closest to our block, however, was the Coconut Grove Park, located east of MacDonald Street, across from Coconut Grove Elementary School. Kids from our section of the Grove had, in recent years, begun spending time at Coconut Grove Park, playing games of pickup football or baseball on the grassy area or basketball on the elementary school courts. Grove Park was two blocks from our homes, much closer and therefore easier to walk to than the six or eight blocks to the closest school park in our own section of town.

There was one afternoon from those days that I'll never forget. Our gang of boys was headed east to the Grove Park for a game of football. Tossing the ball as we walked, we laughed and joked, and talked of the things that young boys talk about. Approaching MacDonald Street, we were holding, not tossing, the football, waiting for a break in the traffic before making our way across the busy highway. Once on the other side of MacDonald Street, we were free to clown around again, and resumed talking while walking down the quiet street that led to the park.

Approaching the park we became aware of a distinct buzzing sound. At first I thought the sound must be from a power tool; perhaps someone was working in his garage or in a nearby yard. As we drew closer to the park, the sound grew louder and less like that of a power tool. Trees obscured our view and we were still unable to see what was going on at the park. Curious now, we stepped up our pace a bit, anxious to see what could be causing the curious sound. Entering the park, our view no longer obstructed by the trees, we saw what to me was a most exciting sight – several boys *flying* model airplanes! The tall, white kids were several years older than we were, high school students I supposed. They confidently and gracefully guided the authentic-looking model planes through great circles in the sky. I had glued together plastic model planes many times, but I had never before seen a real powered flying model airplane. We stood at the edge of the park and watched, marveling at the model airplanes high overhead, making great circles at the end of their long control lines, engine noise filling the normally quiet Miami sky. After a few minutes, an engine sputtered, coughed, and then grew silent. The boy on the control line glided it to a soft landing in the grass. I watched as his partner, another high school-aged boy, refilled the tiny gas tank, attached wires from a large dry-cell battery, turned the propeller with his finger, and brought the engine to life again. He made a small adjustment to the needle valve on the rear of the engine which caused the pitch of the engine sound to change. He made a second slight adjustment and listened to the resulting engine pitch, and then another adjustment, listening again.

Once he had the pitch where he wanted it, indicating that the engine operating rpm was just right, he released the straining airplane. Off it went, into the air, on another incredible journey full of great circles, with a "pilot" not much older than me at the controls.

There were other boys and their friends watching from the sidelines, their models parked on the grass "tarmac." They were waiting their turns for take-off into the model aircraft traffic pattern. I ventured close enough to get a good look at one of the parked models and saw the incredible detail of its construction. I saw the imprint of the ribs underneath the silk skin covering of the wing. I also observed the meticulously painted lines of the color scheme, aircraft identification markings, and the American flag. I could see the model pilot seated in the miniature cockpit underneath a clear plastic canopy, waiting for his turn to take-off.

I was familiar with dry cell batteries, silk skin and ribbed aircraft structures, having read about them in books and seen their pictures in magazines. I had spent many hours sitting, reading model airplane magazines and daydreaming. But here was the real thing.

Junior and I had been firmly taught by our parents that "you do not to touch anything that doesn't belong to you! Especially in stores. People might think you're going to steal something." Also stressed was the admonishment "you might break it and then you'd have to pay for it." In short: look but don't touch. On this occasion though, I was so excited that I disregarded any thoughts of ownership, reached down and picked up one of the model airplanes. I examined it carefully, savoring the way it looked and felt in my hands. Cool to the touch, crisp to the eye, long shapely lines, sturdy construction, lightweight, engine at the ready to whisk it skyward at the beckoning of its human pilot on the end of the control line. My thoughts were interrupted when one of the boys yelled at me from across the field, "Hey! Put that down. Don't pick that up!" It wasn't a mean-spirited command at all, just simply a matter-of-fact request, the same sort of request that I would have made of anyone handling my airplane without permission. I carefully returned the model to its place in the parking area and once again turned my attention skyward. I will never forget the feelings of excitement and disappointment that I felt during those next few moments, as I watched those boys doing what I would have given most anything to do. I felt envy, curiosity, but most of all hopelessness that I would ever have the chance to actually fly an airplane, even in model form.

> **Perhaps it served as fertilizer for a seed planted somewhere in the back of my mind.**

My buddies were intrigued by the airplanes but didn't appear to be affected to the extent that I was. They thought the model airplanes were cool, but "what about our football game?" one of them asked.

I gave one last look skyward and wished for just one small sip of the airborne drink to which I had already become addicted. I then turned back toward our neighborhood explaining, "we can't play football here today."

As we walked home I thought about what had just occurred. I tried to shake my feeling of disappointment but I couldn't. No matter how I tried, I couldn't feel the same after leaving the park as I'd felt when approaching it.

As we walked, something on the ground caught my eye. *There, ahead of us on the ground; what's that in the sand? A wing . . . a model airplane wing!* It was just a simple flat one, made of balsa wood, not nearly as elaborate as those I had seen at the park. It was covered with sand and somewhat scuffed up, as though it had been lying on the ground for some time. Not missing a stride, I picked it up and took it with me. I wondered who had abandoned such a treasure. Didn't he realize that all he needed was the rest of the airplane – a motor, batteries, gasoline, and control lines – to have a complete flying machine? Who could not appreciate the value of such an object? I took that piece of airplane scrap home with me, cleaned it up, sanded it down and repainted it with some of the bright yellow plastic model paint that I had. I chose yellow because that was the color of the Navy model SNJ training airplane in one of my model airplane magazines. I placed that wing on the dresser in the bedroom that Junior and I shared so that I could admire it whenever I entered the room.

I kept that wing for a long time afterward (never obtaining any "rest-of-the-airplane") as a reminder of that day at the Coconut Grove Park. *Just a piece of scrap*, I thought, as I grew older, but in reality much more. Perhaps it served as fertilizer for a seed planted somewhere in the back of my mind, a seed that later blossomed into a dream that took me where very few people have gone, into space.

"Slick! What's going on?" It was the commanding officer, approaching from down the passageway.

"Oh, hi Skipper. I'm just checking things out. That F-14 is one awesome machine. And I just love flying it," I added.

"As it should be," he said. "And you should also love flying it in My Squadron," he added with a loud voice and a big grin.

"I sure do," I replied.

"You going flying tonight?" he asked. I replied, "Yes sir. I'm just waiting for Pink. He and I are going to go out tonight and scare ourselves a bit before going home."

"OK. Fly safe."

"Thanks Skipper. Will do."

I thought to myself, *I'd better hurry on down to Maintenance Control. Perhaps 205 is available for this flight.*

— Chapter 6 —

How Could They Know?

I NEVER SLEPT WELL IN SPACE. LIKE MANY OTHER ASTRONAUTS, I rarely slept continuously, without interruption, through a complete sleep period. And like many other people, I typically don't sleep well here on Earth either. I fall asleep easily enough, remain asleep for a few hours, and then wake up. I usually spend the next few hours awake before eventually falling asleep again. This pattern, common for many people on Earth, was also typical for me in space.

NASA flight surgeons allow astronauts to use prescription sleep aids, but those medications did nothing to improve my sleep in space. Even the current "natural" sleep-inducers, so popular in health food stores, didn't help.

Sleeping in space is different from sleeping on Earth. Astronauts orbit the Earth once every ninety minutes, so they see the Sun set and rise again sixteen times in every twenty-four-hour period. On Earth, we basically go to bed when the Sun sets and we get up when the Sun rises. If we followed this same pattern in space, all we would do is go to bed, get up, go to bed, get up, go to bed . . . We'd never get any work done!

To provide as normal a sleep-wake cycle as possible, our daily schedule in space has awake periods and sleep periods programmed into it. We are scheduled to be awake, working, eating, grooming, and performing other daily activities for sixteen hours. An eight-hour sleep period occurs next. Then another sixteen-hour awake period follows the eight-hour sleep period, and so on throughout the length of the mission.

Sometimes mission timelines require a shift forward or backward in time to accommodate a particular event. For example, if the planned time for a rendezvous with a satellite doesn't fall within a scheduled wake cycle, it might be necessary to

lengthen or shorten earlier wake or sleep cycles over several days prior to the rendezvous. This lengthening or shortening shifts the crew's sleep pattern, so that the rendezvous occurs during a normal wake cycle. Sleep shifting is a common practice for astronauts and is most often accomplished on the ground, before launch, to accommodate the required lift-off time of scheduled on-orbit activities. It was on my second space flight that we launched in mid-afternoon for a sixteen-day mission aboard *Columbia*. We shifted our sleep during the mission to accommodate an early morning landing.

People often ask, "How do you sleep in space?" I jokingly answer, "anyway you want!" After the chuckles fade, I explain that in space astronauts do actually sleep almost anyway and anywhere they want. We sleep either in sleeping bags or in sleep stations.

Sleeping bags are a good choice for a space mission because they are relatively comfortable, light weight, and when not in use, they can be rolled up and stored away leaving valuable cabin space available for other activities.

The sleeping bags we use in space are essentially the same as those used by campers on Earth. A zipper running along its full length opens each sleeping bag. Each bag also has openings through which the occupant's arms and head may protrude once he or she is inside the bag.

In the space shuttle you select a place in the cabin to sleep, typically on the mid-deck, and fasten your sleeping bag at that location using the safety-pin-like clips attached to the edges of the bag. There are numerous attach points located throughout the space shuttle to which the crew can attach sleeping bags or other pieces of equipment. A sleeping bag could just as likely be attached along a bulkhead (wall), along the ceiling, or along the floor in any orientation you might want. It could be attached upside down, face up, face down, head up or down, or anywhere and anyway you wanted, so long as it didn't interfere with the functioning of any cabin equipment. The orientation of the bag doesn't matter because a person's body cannot feel orientation (upside-down or right-side-up, etc.) in space. In spite of what your vision might tell you, one position of the body feels the same as any other.

Opposite page – Which way is up? Floating food packaged aboard Endeavour in 1996 surround STS-72 commander Brian Duffy (holding microphone) and Mission Specialist Dan Barry.

> **It's interesting, and a little bit funny,
> to see astronauts sleeping in space.**

Once you have attached your sleeping bag at your chosen location, you zip open the bag, float yourself down inside of it, and then zip yourself up inside it. The purpose of the sleeping bag is actually to restrain your body, keeping it from floating about the cabin and bumping into walls, floors, or equipment. Your body actually floats around, just a little bit, inside the sleeping bag.

When people fall asleep in space, their muscles completely relax and a curious thing happens – the body bends forward a little at the waist, the head tilts slightly forward, and the arms float upward in front of the body. It's interesting, and a little bit funny, to see astronauts sleeping in space.

Many astronauts prefer sleep stations. Sleep stations are individual compartments stacked vertically and oriented horizontally along a space shuttle mid-deck bulkhead. Sleep stations don't provide much room but they do allow for more privacy than sleeping bags. I suspect that I had about as much room in my sleep station as Count Dracula has in his coffin! Each sleep station is accessed through a small sliding door along its side and contains a sleeping bag inside of its hard compartment shell. Each station also contains vents through which fresh air is circulated, an overhead reading lamp, and a soft cloth compartment for storage of items such as a tape or CD player.

Because there is no conventional day or night in orbit, we simulate nighttime by placing window shades over the space shuttle's windows. The shades keep out the light from the constantly rising and setting Sun. We also turn off the lights, darkening the cabin and aiding in the illusion of nighttime. After inducing darkness we enter our individual sleeping "quarters" and settle down to sleep for the next eight hours.

As weird as it feels to live in space, I enjoyed every moment of it. I enjoyed the long busy days (awake periods) and the peaceful nights (sleep periods). I fondly remember "evenings" when I would say good night to my crew mates then float up towards my station at the top of the cabin. I would then rotate my body into a horizontal position so that I could float, feet first, into my sleep station. I'd turn on the sleep station overhead reading lamp and slide the door closed. I had become accustomed, and thus oblivious, to the constant rushing sound of air flowing in through the air vent, and the high-pitched whine of the shuttle cabin air cleaner; this sound penetrated my sleep station. These ambient noises, which might keep most people awake, almost hummed me to sleep. I usually closed the air vent a little for comfort because of the high volume of cold air entering the compartment. I was careful not to close the vent completely (not that the design would allow me to do so) and completely shut off the inflow of fresh air, because of the potential for buildup of carbon dioxide. In short pants, tee shirt, and athletic socks, I would zip myself up inside the station sleeping bag.

STS-72 Mission Specialist Winston E. Scott prepares to enter the Space shuttle *Endeavour* at Launch Pad 39B, as white room closeout crew members Mike Mangione (foreground) and Jim Davis assist him.

Now for my music, I thought at the beginning of a typical sleep period. I placed headphones over my ears while the attached tape player floated in front of me. Reaching for the small, soft bag containing my collection of tapes I thought, *perhaps tonight, Miles Davis' In A Silent Way. I love it. It sounds like . . . outer space!* The continuous low-pitched musical hum of the bass, with pretty upper register harmonies, sound peaceful, like a ship in a quiet orbit. The higher-pitched embellishments occurring here and there over the low, soft foundation remind me of the bright stars against a dark sky. Then, softly, Miles begins to play. His beautiful trumpet melody seduces me and I'm totally absorbed in this experience in space, and I'm forever changed by all that's happening to me here in this orbital heaven. On and on, moment by moment, I lie there, floating in outer space, rising and falling with the music. How could he have known? Miles Davis, who had never flown in space, had written what is to my mind is the one piece of music that most describes space. Soothed, I close my eyes, relax, and drift off to sleep.

Reflections From Earth Orbit

I awaken a few hours later and take a moment to orient myself. I feel as though I'm lying face up, but when I slide open my sleep compartment door I realize that I've floated into a different position. Despite the face-up position I *feel* like I'm in, I'm actually facing downward. Miles is still playing in my headset, the tape having played to the end, auto-reversed itself, and then started again. I've learned to identify each button on the tape player by its location and can actuate the right one without turning on the reading lamp. I press the Stop button and remove the headphones from my ears. I secure the tape player inside the compartment to keep it from floating away, and then I weightlessly, silently float myself to the WCS (Waste Collection Station) compartment for a bathroom visit. As I float through the crowded cabin, gently touching here and there, I'm careful not to bump or bang into anything. I don't want to disturb my crew mates any more than absolutely necessary.

Shortly, I'm back in my sleep compartment with the sleeping bag zipped up around my shoulders. With my earphones again in place, I prepare to enjoy more music until the magic of sleep returns. I must turn on the reading lamp at this point and select a different tape. I remove the tape player from its soft pouch and release it into mid-air. It drifts away, down into the station towards my feet, as far as the headphone cord will allow. I can't see it, but I know it's there, in mid-air, somewhere in the sleep station. I give a gentle tug on the headphone cable and the player floats back to me. I very gently stabilize it in mid-air, coaxing it to stay put. As it floats there in front of me, it bobs about just a bit, giving a little tug on the wires to the headphones against my ears. Now I switch on the lamp and, *Let's see. This time, how about Quincy Jones' Walking In Space? How could he know?* The "Q" has never flown in space! *How could he know?*

I've never slept well in space. I do recall, however, always being asleep at the end of the sleep period and having to be awakened by our alarm, and the wake-up music transmitted from Mission Control. I've always felt alert on awakening, and ready to go with the day's upcoming activities. I remember the excitement of each new day and how important it was to enjoy every moment of each day. I have no memories of feeling tired or sleepy in orbit.

I've never slept well in space. If I should get to fly in space again, I most likely will experience the same sleep patterns that I have for most of my life. Though I might not sleep well, I will most certainly have rested well, with a little help from my friends Miles and the "Q." *How could they know?*

Opposite page – Orbital sunrise during STS-72 as viewed from the flight deck of *Endeavour*.

— Chapter 7 —

The Best Seat in the House

"WELL, THAT MEAL WILL SUSTAIN LIFE," I REMARKED somewhat sarcastically as I finished my sandwich and chicken-flavored rice. It was mealtime in orbit and I was on the crowded mid-deck of the space shuttle *Endeavour* with my five crew mates. In addition to a sandwich and rice, I had eaten green beans and sipped a bagged apple drink through a plastic straw. I was careful to close the clip on the straw after drinking so that the remaining liquid wouldn't float up and out. All liquids form globules, balls, in the weightlessness of space. Had I not closed the straw on my drink bag, I would have soon had apple juice balls of various sizes floating throughout the mid-deck. The juice balls would have slowly moved with the circulating air currents throughout the ship until they splattered themselves on a bulkhead (wall), or perhaps on a piece of machinery. Electrical machinery, of course, is not compatible with apple juice. Also, it would have then been necessary for me to spend precious time cleaning up the sticky mess with a damp towel.

My sandwich consisted of a chicken patty wrapped inside a Mexican tortilla. It was flavored with ketchup from one of the small plastic "to go" condiment packets that we carried in our food bins. Tortillas are the standard breadstuff of space shuttle astronauts. They're thin, fold well, keep well, and taste good! The chicken patty was a military-style meals-ready-to-eat (MRE) morsel that I heated inside our convection oven. It was pre-cooked for NASA at some preparation facility and stored inside a sealed, airtight plastic bag. I simply heated the bagged patty before opening it and making the sandwich. Eating a chicken patty sandwich in weightlessness is so much easier than attempting to eat the same patty with a knife and fork. It's difficult cutting chicken – or any food – with a knife while in space; the food simply won't remain in place on the plate when you try to cut it. If you are somehow able to cut food into pieces, then each piece is likely to go floating up and away.

> # I suppose current space food is tasty . . .
> # It certainly isn't as good as home cooking.

The green beans, like many of our space foods, were dehydrated and packaged inside a dish shaped cellophane container. This container was manufactured with a small portal to allow for the injection of water for re-hydration. I wanted hot beans, of course; therefore I injected hot water from the shuttle's galley re-hydration station. After injecting the hot water, I waited the prescribed amount of time (3-5 minutes for green beans) and then seasoned the beans by squirting small amounts of liquid salt and liquid pepper into the packet. It's impossible to use conventional salt and pepper shakers in space. Attempting to do so would result in the salt and pepper floating all over the space shuttle and certainly not falling onto the food. On Earth, its gravity that causes the salt and pepper to fall onto your food.

My rice had also been rehydrated, same as the beans. The rice was adequately seasoned already and therefore required no additional liquid salt or pepper.

Eating in space is a delicate and challenging activity – no matter how careful you are, it is inevitable that some of the food will float away from you. When this happens, a game of "catch and eat" ensues with the astronaut attempting to catch the floating food with a utensil before depositing it into the mouth. An experienced astronaut simply floats on a trajectory designed to intercept the escaping morsel and catch it in his open mouth. Being in space can be such fun! We astronauts play many games in zero gravity and many of these games involve food.

I suppose current space food is tasty when compared with that toothpaste-tube stuff of the early days of space flight. It certainly isn't as good as home cooking!

Let's see, how about chocolate pudding for dessert, I thought as I collected my empty food containers. *Or maybe, fruit cocktail!*

One by one, I reached for each empty container. They had been fastened with Velcro patches onto the shuttle's bulkheads or mid-deck locker doors so that they could be collected and disposed of at the end of the meal. I squeezed the packets together and compacted them in my hands, and then I slid my feet out of the floor-mounted foot-loop restraints that I used to hold myself stationary on the deck. I gently pushed off and drifted upwards towards the upper food lockers. I moved

very slowly upward and then stopped myself with my outstretched hand on the overhead (ceiling). Suspended now in mid-air by the weightlessness of space, I placed the empty food containers inside the oblong, clear plastic garbage bag. I squeezed and compacted the bag, and folded the top over several times, then I resealed it with a large clamp, and restored it to its holding place. My body movements while handling the garbage pack had caused me to drift out of my desired position. My feet had floated up and out in front of me, and I was slowly beginning to drift into a head-down position. I put out my hand and gently touched the overhead and returned myself to my previous "feet-down" orientation and my previous spot in mid-air, in front of the upper food lockers. While still floating there, I opened one of the lockers and surveyed the snacks inside. I like chocolate, so the choice of pudding would be easy. On the other hand, *fruit cocktail and vanilla cookies would taste pretty good right about now*, I thought. It's interesting that the pudding and fruit cocktail are packaged in the same small individual cans that anyone can purchase here on Earth at any local grocery store.

> ## I have always imagined that this must be how Superman feels when he comes in for a landing.

Reaching inside the food drawer, I carefully placed my hand underneath the cover netting that held the foodstuffs in place and selected a small, round tin of fruit cocktail. I had learned early on to be careful when removing food or anything else from its drawer, so as not to inadvertently send the other items in the drawer floating out into the cabin. If I wasn't careful, I'd find myself chasing food items all over the space shuttle mid-deck.

A gentle push on the overhead sent me moving, ever so slowly and comfortably, to the floor again. I have always imagined that this must be how Superman feels when he comes in for a landing. Transitioning from high-speed forward flight, rotating his body so that his feet are beneath him, he descends to Earth in a controlled and graceful manner. As I reached the deck, I tucked my right foot and then my left inside foot loops again and pulled myself down into the slightly crouching position, similar to sitting, that is so comfortable in zero gravity. I tucked in my red cape, pulled on a jacket over my T-shirt – you know, the one with the bright S on it – and returned my attention to my dessert.

Opposite page – The crew of STS-87: (a) Kevin R. Kregel, Commander; (b) Steven W. Lindsey, Pilot; (c) Winston E. Scott, Mission Specialist; (d) Kulpana Chawla, Mission Specialist; (e) Takao Doi, (NASDA) Mission Specialist; (f) Leonid K. Kadenyuk, (NSAU) Payload Specialist.

STS-87 Mission Specialist Winston Scott poses in his orange launch and entry spacesuit with NASA suit technicians at Launch Pad 39B during Terminal Countdown Demonstration Test activities.

I rotated the round fruit cocktail tin in my hands so that I could pull the pop-top, and as I did so I noticed a glint of light reflected from its shiny surface. It was a mild glint in the soft shuttle lights, but noticeable never the less. It's fascinating how seemingly unimportant things can remind us of events that happened in earlier times and places in our lives. A faint smell of pine might perhaps remind a man of his first camping trip as a boy. The sound of a particular door chime might remind a woman of the first time she, as a young girl, was picked up by a boy for a date. The glint of light from that fruit cocktail can reminded me of the first time that I had seen a similar phenomenon. It was the first time I saw the Sun reflecting off the face of a shiny coin, a quarter. It reminded me of the days in my early childhood when a quarter was such an important part of my day. Those were the days when, as a brand new first grade pupil in 1956, Mama would give me a quarter for lunch each day before taking me to school.

George Washington Carver Elementary School sat just on the edge of the "colored" section of Coconut Grove, where Grand Avenue intersected highway US 1. (US 1 is still known as the Dixie Highway.) Carver actually served "colored students," as we were known in the 50's, from kindergarten through twelfth grade. It was a warm and friendly school, as I would grow to learn. The elementary section bordered Grand Avenue. The Junior / Senior high school section was located at the end of what seemed (to my first grade eyes) to be miles of raised corridors. The elementary school classrooms were evenly spaced along these corridors.

My sister LaVerne was in junior high school at the time that I began first grade. She was a very good student who had started first grade at the age of four. She was my big sister and looked out for me these first days in school.

Mama had a daily routine worked out for us. She dropped LaVerne and I off at the school. After depositing me on the elementary playground to wait for my bell, LaVerne proceeded to her classes in the high school section of the huge campus. I remained on the playground playing until the elementary school bell rang, then I would go to my classroom. I knew where my classroom was on my first day of school because I had visited there on registration day.

While playing on the playground on that first day of school, a boy, older and bigger than me, came up and said, "Hi." I don't recall his name now, but I do remember him saying that he was in sixth grade. I told him I was only in first grade. We played together in the park, riding the merry-go-round, swings, and the seesaw. He was friendly and helpful, giving me a hand on the rides that were a little big for me to handle on my own.

When the bell rang and I was ready to go to my classroom, I suddenly realized I had my little notebook, my long wooden pencil, but no quarter. The pocket in which I had put it was empty.

> **I was not eating lunch at school because I never had any money.**

I told my new friend that I could not find my quarter and he agreed to help me look for it. We searched the playground, looking in the grass and the sand for my quarter, but didn't find it. The bell had rung and I knew I had to go. I went to my class and began what was to be a long, trying first day at school.

I don't remember much else about that first day, but I do recall that having no quarter meant having no lunch! While the class ate in the big cafeteria, I put my head down on the table where we sat and fell asleep. When I woke up, the class was gone. I wasn't worried or concerned at all. I knew where my classroom was located and I simply got up from the table, went to the classroom, and took my seat. The class had only arrived back in the room a moment before me and I supposed the teacher had not had a chance to miss me. Still, I could not understand for sure how and why my teacher and the rest of the class had left me in the cafeteria.

The rest of that day is a blur to me now, but it's easy to recall what happened during the subsequent days. Those next few days were at first a variation on the same theme. I was dropped off at the school playground where I would meet my friend, play, and lose my lunch money. He would help me search for it to no avail, and I would go to class and spend the day in school with no lunch. My teacher did notice that I wasn't eating and asked me about my lunch money. I remember telling her simply, "I lost it."

Neighborhoods were different in those days. They were true communities where everyone knew everyone else and looked out for one another, at least more so than it seems today. Teachers, preachers, merchants, and parents all possessed that sense of ownership for the whole community and showed concern for one another's children. As was the norm then, my teacher paid for my lunch a couple of times but then contacted my mother and let her know that I was not eating lunch

Opposite page – STS-87 photography – (a) Ganges Delta, India: A glacier at about 22,100 feet in the Himalayas is the source of the Ganges river. Hundreds of miles later and joined by other tributaries the Ganges delta enters the Bay of Bengal. The delta, at 200 miles wide (320 km) is one of the most fertile and densely populated regions of the world. (b) The Sinai Peninsula, Egypt: Viewing the mountainous region of the St. Katherine Protectorate from space, a gray patch is visible on top of a heavily visited mountain.

Reflections From Earth Orbit

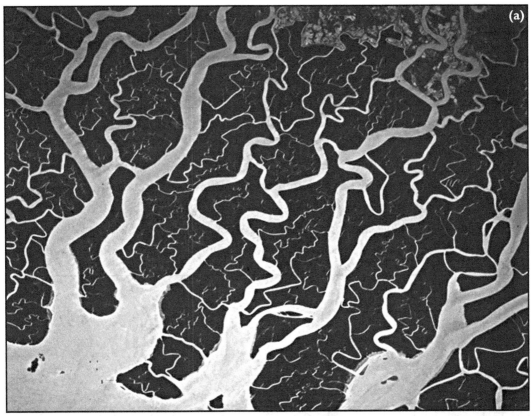

at school because I never had any money. I can surmise that there was much more to that conversation, but I was not privy to it. I do know for sure that it took place, because Mama talked with me about the situation afterward. I told her about playing on the park with my friend and losing my quarter. I recall her saying something about him, my friend, taking my quarter. I defended him however, saying again and again that it was I who had lost my lunch money!

Mama was faced with a problem. How could she give a six-year-old boy a quarter, allow him to play on a playground and have him retain that quarter? If Mama had been schooled in materials science she might have been the developer of Velcro. She would have installed it on my trouser pocket to keep the pocket tightly closed and the quarter in. If Mama had been schooled in textiles, she might have

STS-87 astronaut crew members prepare to fly back to Johnson Space Center in Houston after participating in the Crew Equipment Integration Test (CEIT) at Kennedy Space Center. The CEIT gave the astronauts an opportunity to get a hands-on look at the payloads with which they would be working in orbit.

designed a fancy wallet full of pockets and flaps and zippers to fit inside of the pockets and flaps and zippers on my trousers, to ensure the security of my quarter. If she had possessed a degree in advanced economics, she might have eliminated the need for me to carry the quarter altogether, by introducing the first debit account system. But Mama only had what all other mothers possess. She had that special ingenuity that comes from the need of mothers to protect their children.

Today's NASA would be proud of Mama with her "better, faster, cheaper" solution to my economic problem. The day after my teacher spoke with her, as she prepared me for school, she took a pocket handkerchief, rolled my quarter in one corner, then tied that corner, quarter and all, into a knot. The knot was tight and small enough to hold the quarter securely in the handkerchief, but large enough for my six-year-old fingers to loosen when the time came to buy lunch. She then used a safety pin to attach the handkerchief, quarter and all, to the inside of my pants pocket. I was *ready* now! No one, not even my brave and strong TV hero Sky King (not that he would do such a thing) could have separated me from my quarter, at least not without me knowing it. I ate lunch that day and every day thereafter. It's curious though; my playground friend suddenly stopped playing with me in the mornings.

Mothers! Perhaps the most creative creatures on Earth. I have many wonderful memories of my mother. She could be firm, even stern, yet she always had a joke or a riddle to share. She could keep us laughing with her great sense of humor. Her handling of my lunch money losses stands out in my mind as testament to loving mothers everywhere.

Mama didn't get to see me fly in space. She didn't even get to visit our family when we moved to Houston, Texas after my astronaut selection. I reported to Johnson Space Center in Houston in August 1992 and I flew my first space shuttle mission in January 1996. Mama died in July 1993. She would have been so proud and excited to be at the Cape, with other family and friends, for my first launch. Mama did get to know all four of her grandchildren however, my two and LaVerne's two. Her grandchildren all called her "Nana."

Mama's oldest granddaughter, my niece, Jounice, was a young newspaper reporter at the time of my first space flight in 1996. Jounice wrote a series of articles for her newspaper about her uncle, the astronaut. The articles followed me, our family, and selected friends as I progressed through training towards my first mission into space. The final article of the series was published after my launch. In that story LaVerne, Jounice's mom, full of emotion, is quoted as wistfully saying, "I wish Mama were here to see Winston's space shuttle launch. She wanted so badly to be here for this." Through her customary big bright smile, Jounice replied to her mother, "Don't worry mom, Nana has the best seat in the house."

"Hey Winston, you going to eat fruit cocktail or chocolate pudding" asked my crew mate Leroy? I replied, "I guess I'll eat this fruit cocktail." Then I thought, *Mama would have wanted it that way, and she's probably watching!*

On board space shuttle *Columbia*'s (STS-87) first ever Extravehicular Activity (EVA), astronaut Takao Doi works with a 156-pound crane carried on board for the first time. The crane's inclusion and the work with it are part of a continuing preparation effort for future work on the International Space Station (ISS). The ongoing project allows for evaluation of tools and operating methods to be applied to the construction of the Space Station. This crane device is designed to aid future space walkers in transporting Orbital Replacement Units (ORU), with a mass up to 600 pounds (like the simulated battery pictured here), from translating carts on the exterior of ISS to various work sites on the truss structure.

— Chapter 8 —

Night Pass

YOU'VE NEVER SEEN DARK UNTIL YOU'VE SEEN THE DARKness of outer space. It's a darkness unlike the frightening kind you saw when you were a child. That darkness enveloped your bedroom and caused you to pull the covers over your head to hide from the monsters under the bed. And outer space is not made up of the threatening kind of darkness that you see during a particularly violent thunderstorm, a hurricane, or even a tornado. There's no violent rain, hail, thunder, or lightning present. The darkness of space, at least in low-Earth orbit, is peaceful. If there's any such thing as a calm and beautiful darkness, then it is there, in space.

You might think that the features of the universe are visible at all times in orbit. After all, you're in space. You might think that all you have to do is look out the window and there they are: the stars, the planets, the Sun and the Moon. In reality, what you see in space depends upon where you are along your orbital path. On the bright daylight side of orbit, the Earth can be seen clearly but the other planets and the stars are not visible. On the dark side of an orbit, the features of the Earth are not visible but other planets, the stars, and the Moon can be seen clearly. In order to see these heavenly bodies, however, lighting conditions must be right. The ambient light must be reduced sufficiently so as to not obscure these objects. In the space shuttle, viewing is done from the upper flight deck, where viewing windows are located. The lights on the flight deck must be turned off. We must also curtain off the flight deck from the mid-deck to prevent mid-deck light from filtering up and preventing our view of space. The shuttle's external payload bay lights must be extinguished also. Once these things are done, the night sights of our solar system are available for us to behold.

> # The sight of it will live in my mind's eye for the rest of my life.

The night sky viewed from the Earth's surface looks like a dark tapestry with the stars appearing as specs of white light. With few exceptions, the stars all look the same – they exhibit no depth or color. When viewed from space, stars take on a more three-dimensional appearance and some stars actually appear closer to us than others. Some stars look larger and brighter than others, and all of them are magnificent. I vividly recall night passes where constellations, easily recognizable, filled the space shuttle's windows. Orion, a giant of the night sky, stood watch over our part of the universe. Leo the lion, the big dipper and the little dipper were clearly visible. One of the most magnificent images indelibly etched in my mind is one that I saw from the flight deck of the space shuttle *Columbia*. We were on *Columbia*'s aft flight deck, gazing out the rear windows when a beautiful sight began to materialize. Off to the right we could see Mercury and Venus. Earth, of course, was just below us. We could see the three planets simultaneously, and as our ship moved through space, at close to eighteen thousand miles per hour, the three planets appeared to be moving into a straight-line formation. They moved toward one another very slowly until, for a brief instant, I could see all three planets at once – Mercury, Venus and Earth – in what appeared to be a straight line. At that same instant, out to the left of me, I could see the Moon. This formation lasted only a moment, but the sight of it will live in my mind's eye for the rest of my life.

It's fun to watch the space shuttle jets fire in darkness. There's no air in space – at least, not enough to allow the use of conventional airplane-type flight controls. The space shuttle is maneuvered from one attitude to another in this airless environment through the use of a system of RCS (Reaction Control System) jets. The bright plume of fire from the RCS is invisible during daylight but becomes brilliant and beautiful against the dark backdrop of a nighttime space sky. These jets typically fire short, rapid bursts of energy, more resembling cannon fire than jet exhaust.

Another splendid spectacle occurs when an occasional glow appears around the tail of the space shuttle, caused by the presence of electrical charges from ionized gases in space – a sort of St. Elmo's fire in space! A most impressive sight.

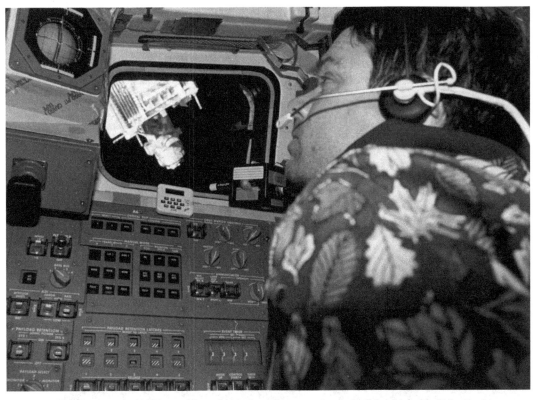
STS-87 commander Kevin Kregel watched Spartan through *Columbia*'s aft flight deck.

Although you can't see specific features, the Earth's surface, dotted by city lights, is impressive when seen from space at night. The whole Earth was caught up in the effects of El Niño ocean-atmosphere disruption in 1996. The resulting wild fires in Australia, which consumed thousands of acres of land, were clearly visible from space during our night passes.

I vividly remember looking down on the night sky and watching lightning ripple through the dark thunderclouds miles below me. I also recall another particularly awesome sight. I happened to be looking down at the dark Earth surface at just the right moment when, suddenly and silently, a shooting star passed beneath me. A distant bright dot with a fiery tail trailing behind, the flaming meteorite lasted only a moment, and then disappeared, burned up in the atmosphere.

I remember approaching the night-covered west coast of North America. I couldn't see the coastline specifically, but I could identify the California cities by their location. Out of the right space shuttle window was a bright spot of city lights. I surmised that that was San Diego. My wife and I had spent my first eight

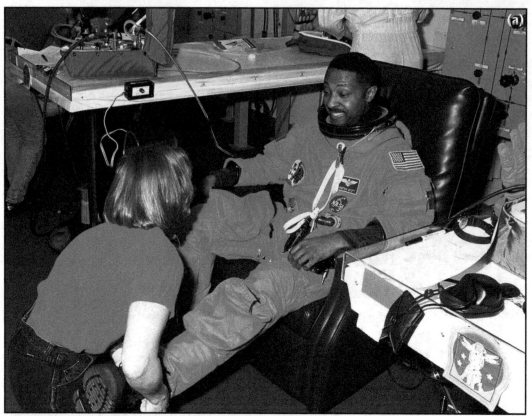

years in the Navy, after my flight training, living in San Diego. A little to the left of that cluster of lights was another one – *that must be Los Angeles*, I thought. We never lived there, but we visited L.A. many times. My wife's uncle lived there. Our first visit to a Disney theme park was in Los Angeles. A little further on, to the left of L.A., was another cluster of lights that must have been San Francisco. We'd had fun times visiting San Francisco when we lived in Monterey, California, where I had attended graduate school. I was thoroughly engrossed in my sight seeing. While over the middle of North America, over Texas, I saw a large cluster of lights that I knew was Houston. A thin sliver of light wound its way to the south, to a smaller cluster. That sliver was the highway and bridge leading to Galveston, the southernmost light cluster. The most impressive sight was over the east coast of North America. I saw a large area of light that I recognized as Miami. A thin line of lights that was the causeway led to the cluster of lights that I knew was Key West. What was absolutely breathtaking was the view of the entire state of Florida, with its unique shape, outlined in city lights. On a grander scale, I could see all the way to Puerto Rico out of the right window, and out of the left window I could see Boston. It was so beautiful, so impressive.

If the stars were notes, the sky would play a symphony.

If the stars were notes, the sky would play a symphony. If the oceans were paints, the Earth would produce a masterpiece. No notes, no paints, no words can convey the beauty of what I saw from the flight deck of *Endeavour* that night. I saw the magnificence of our universe on this night pass and the others that followed. That beautiful sight will be forever etched in my mind.

Opposite page – STS-87 Mission Specialists (a) Winston Scott (NASA) and (b) Takao Doi, (National Space Development Agency of Japan) don their launch and entry suits in the Operations and Checkout Building. Scott and Doi performed an extravehicular activity spacewalk during the mission.

— *Chapter 9* —

We've Got to Have Arithmetic

"**J**ACKSONVILLE CENTER. NASA 901 CHECKING IN AT FLIGHT Level three-nine-zero."

I reduced the power on the two J-85 turbojet engines that propelled my T-38 jet aircraft. Once level at a cruising altitude of thirty-nine thousand feet, I allowed the aircraft to accelerate to cruise speed and then set cruise power. *Let's see*, I thought; *Fuel flow: approximately 750 pounds per hour, per side ought to give me Mach zero point nine. Nine-tenths of the speed of sound at thirty-nine thousand feet. Pretty good!*

I could see the coast of the Florida panhandle off to my right side as I made the familiar trip from Kennedy Space Center – The Cape – to Houston, Texas. Although we fly the space shuttle from the Cape in Florida, NASA astronauts live and work in Houston. The Gulf of Mexico below me looked calm. It almost made me believe I could enjoy boating as much as I was enjoying myself flying at flight level three-nine-zero. But I wasn't fooled! The bright, blue water was rimmed to my right by the white sand of the Florida beaches. There were sailboats below, quiet and small, just specks on the blue carpet of ocean. And pretty soon, I knew, I'd be able to see tiny spots on the water that were oil rig platforms. And as I approached Houston, oil tankers and fishing trawlers would be plentiful; I could count them if I wanted to, as they crawled to their destinations.

The sky was beautiful, clear with only an occasional white puffy cloud. *I'm glad I was able to take off as early as I did this afternoon*, I thought. If I had taken off any later, the hot southern Sun would have been lower in the western sky and shining directly into my eyes. Even through my colored helmet visor, the direct sunlight is very uncomfortable.

Two T-38 jets departing Kennedy Space Center for their home at the Johnson Space Center in Houston, Texas.

Hm-m-m, I thought. *If I experience an aircraft problem now, I can divert into Eglin Air Force Base. Eglin's weather is good.*

The weather forecaster predicted winds at thirty-nine thousand feet to be out of two-seven-zero at fifty knots. *That's fifty knots on the nose. That'll slow me down, but I should be in good shape fuel-wise. I'll check the computer and calculate it myself, just for backup.*

"NASA 901, this is Jax Center, radar contact at flight level three-nine-zero. Good afternoon."

"Good afternoon Jax," I replied.

This image highlights coastal areas of four states along the Gulf of Mexico: Louisiana, Mississippi, Alabama and part of the Florida panhandle. The image represents an area of 345 kilometers x 315 kilometers.

I'd flown this flight profile so often that I regularly joked that the aircraft could find its own way to and from the Cape. I could fall asleep and the aircraft, like a faithful guide dog, would take me to the Cape if I'd asked it to. Or maybe it was more like a homing pigeon, always returning to Houston when I was away and wanted to go home. This, of course, wasn't true, but I did fly this route so often that it would become a little boring if I didn't keep my mind occupied. I was alone on this flight; there was no one in the back seat to keep me company. *Mustn't let my mind wander too much while flying*, I told myself.

Let me double-check my ground speed, I thought. *OK. I'm receiving the signal from Leeville* (a ground station directly on my flight path). *That simplifies my backup calculations. I'll start a timer then check it again in a moment. A little basic math will help me determine whether or not the computer is accurate.*

"I'm a skinny piece of leather, but I'm well put together."

I was always a pretty good math student, even way back in elementary school. I vividly recall the day that a teacher unfamiliar to us entered our fifth-grade classroom, spoke with our regular classroom teacher, and then called out my name along with the names of several other students. We were to accompany this other teacher, Mr. Franklin Clark, to a different classroom.

Mr. Clark led us outside and along the second-story corridor to his classroom. There, he explained to us that we were selected to participate in an accelerated reading and arithmetic program. While our classmates were having math and reading lessons with our regular classroom teacher, our group would have math and reading with Mr. Clark.

Mr. Clark was an unusually animated individual. He walked with a fast gait, spoke with great emphasis, and even sang with us with much enthusiasm. Though somewhat shorter than most of the other male teachers at the school, he possessed a larger than average energy level. He was always friendly, almost bubbly, and supportive to his students. Mr. Clark, however, demanded discipline and respect. We mimicked him behind his back (all students mimic their teachers) but we liked and respected him. He would sometimes "threaten" us with having to settle any classroom infractions personally with him. He would assume this really serious facial expression and proclaim, "I'm a skinny piece of leather, but I'm well put together." It was difficult for us to conceal our snickering, since he didn't seem nearly as menacing as some of the other male teachers. We were careful, however, to exhibit good behavior while in his presence.

Mr. Clark guided us as we explored the concepts of time and speed. Unbeknownst to us at the time, we were also reinforcing our basic arithmetic skills. I recall discussing the concept of a light-year. We actually calculated by hand (there were no hand-held calculators or classroom computers in those days) the distance that light traveled in one year, a light-year. We raced against one another to see who could arrive at the answer first, yelling out our answers, tremendously

long strings of numbers. Our answers were probably not numerically correct, but accuracy wasn't important within that particular context. What was important was the thought processes involved in understanding the concept of a light-year, understanding why this concept is important, and understanding the mechanics of making the light-year calculations.

Oh, I almost forgot! Must calculate my ground speed. Don't want to blindly trust the FMS (Flight Management System) computer. Let's see . . . Leeville is at seventy-eight nautical miles now. That means I traveled approximately eight nautical miles in ten minutes, therefore I must be making four-hundred-and-eighty knots ground speed. The FMS computer says four hundred eighty six knots. Good, the two answers are very close. Four-hundred-and-eighty knots, with a one point one five multiplication factor, is approximately equal to five-hundred-and-fifty miles per hour. I smile at myself. Making such mental calculations is not necessary; It's not required by any regulations. In fact, the computer is almost never wrong. These calculations, however, give me that "warm and fuzzy feeling" as we say, when the answers match those of the computer. (Besides, I've read where performing mental gymnastics exercises the mind and wards off Alzheimer's. Hope it's true).

"NASA 901, Jax. You've got traffic at two o'clock, one-zero miles, opposite direction, level."

"Roger that Jax. Nine-zero-one's looking," I replied.

The traffic was another aircraft just slightly to the right of my aircraft's nose. Flight controllers will often give pilots this type of warning call so that they might see each other's aircraft and augment radar separation with visual separation. I called to Jax, "901 has the traffic in sight." I could see a big airliner passing down and to the right of my flight path. The blue and white color scheme, with a delta pattern on the tail, allowed me to readily identify the passing aircraft. It was a Delta Airlines jet, eastbound, probably heading to Atlanta, Delta's hub.

It's nice and quiet here at altitude, and a smooth ride to boot, I thought. I hate it when there are thunderstorms all over. It really adds an element of difficulty to

Opposite page – Like a rising sun lighting up the afternoon sky, the space shuttle *Columbia* soars from Launch Pad 39B at 2:46:00 p.m. EST, November 19, on the fourth flight of the United States Microgravity Payload and Spartan-201 satellite.

the flight when you have to duck and dodge thunderclouds, with the associated hail and lightning, not to mention the fact that you can't see where you're going and have to depend one-hundred percent on your gauges. The on-board color weather radarscope is clear today, to my delight. I'm delighted also for the smooth, uninterrupted whine of the flawlessly operating engines. The fuel balance is good: less than a five hundred pound split between the right and left tanks. The distribution of fuel inside the aircraft tanks is critical to maintaining the proper balance and control of the aircraft. If the fuel split grew any larger, I'd have to manually balance the fuel. No big deal. I scan the engine instruments and note that the engine oil temperature, oil pressure, and hydraulics systems all look good. The seeming tranquillity of the situation belies the fact that I'm slicing through the atmosphere at close to the speed of sound. It feels good.

Mr. Clark introduced to us the concept of binary numbers. He talked to us about how our decimal system is based on the value of ten. The binary system, on the other hand, is based on the number two. I didn't understand until much later just how the binary system is utilized, but we sure did have fun that day in Mr. Clark's class, counting in base two.

Mr. Clark taught us many things. He constantly admonished us to "answer all questions, either verbally or written, with a complete sentence." In his class we read books; we read and wrote poetry; we sang songs and talked about current events; we followed the events surrounding the selection of the original seven Mercury astronauts and their early space flights. The most valuable lesson he taught us was the importance of perseverance and hard work. He had us memorize a favorite poem, by an anonymous author, that I can recall even now, some forty years later.

"If a task is once begun
Never leave it 'til it's done.
Be the labor, great or small,
Do it well, or not at all."

"*Be* the labor." I like that statement. Don't just do it, *be* it! – embrace it, become a part of what you're doing, and do it to the best of your ability. This is a valuable lesson for all of us.

Let's take this period and play softball in the park.

"NASA 901, Jax. Contact Houston Center on frequency one-three-two-one-seven."

"Roger Jax. 901 switching to Houston, one-three-two-one-seven."

OK, I thought, *select VHF Comm One* (radio one). *There's the rectangular cursor around the frequency window. Now select frequency 132.17, then toggle this freq into the active frequency window. Pause for a few moments to be sure I don't interrupt a radio conversation already in progress, and then*, "Houston Center, NASA 901, with you at flight level three-nine-zero."

After a moment the reply comes, "NASA 901, Houston Center radar contact."

Then there was the one particular day in Mr. Clark's room. He was busy, as usual, arranging materials for class, placing papers here, a book there, continuously moving. While several of us were gathering, preparing for class, we made light conversation with Mr. Clark. Only half seriously I said, "Mr. Clark, let's not have arithmetic today. Let's take this period and play softball in the park."

Mr. Clark did not pause from his pre-class preparations. He did not even look at me when he answered, but he did answer, softly, with a hint of a smile, and with a conviction in his voice that I can still hear today, "We've got to have arithmetic," he said. Then he repeated for emphasis, *"We've Got To Have Arithmetic."*

"NASA 901, descend to and maintain ten thousand feet. Cross twenty-five miles east of Trinity, at ten thousand."

"Roger that Houston. Nine-zero-one is out of flight level three-nine-zero for ten thousand. Will cross twenty-five miles east of Trinity at ten thousand," I responded.

Doing mental math again, I thought, *Trinity is seventy-five miles from here. At eight miles per minute, it will take me roughly nine minutes. I need to lose nineteen thousand feet* (approximately) *over the nine minutes. A descent rate of two thousand feet per minute is a good figure to shoot for. Close enough for government work*, I quipped to myself.

I've seen Mr. Clark only twice since elementary school. I didn't have a chance to talk with him on either occasion. If I ever do see him again, if I get just a moment, I will remind him of that day in his classroom, when a certain fifth-grade future astronaut wanted to play softball rather than have arithmetic. I'd tell him how glad I am that we had arithmetic on that day, and on every other school day. I'll simply say, "Mr. Clark, I'm so glad we had arithmetic." Then maybe for emphasis I'll say it again. *"Mr. Clark, I'm Really Glad We Had Arithmetic."*

STS-87 photography – Gulf of Aqaba, Egypt: Significant changes in coastal land-use through construction of resort areas and major roads are evident in this photograph.

— Chapter 10 —

It's About Time

I'VE SOLVED EINSTEIN'S EQUATIONS (SIMPLIFIED OF COURSE) FOR relativistic time dilation. Dr. Einstein showed, at least from a mathematical standpoint, that a body traveling at a velocity close to light-speed would appear to experience time dilation – time passing more slowly than experienced by a body at normal ("non-relativistic") velocity. This would mean that while people were traveling at high speeds (like astronauts) they would age more slowly than people who remained at normal speeds on the ground. Would this mean that, according to Einstein, I am younger than I would have been had I not flown in space?

Time is such an interesting phenomenon, yet most of us take it for granted. It is said that we spend too much time in front of the television set, or too little time exercising our minds and bodies. Our parents consistently reminded us to not "waste our time" and that "time waits for no one." Maybe we procrastinate a bit too much and find ourselves in a panic when we've arrived at the proverbial "last minute," and what we thought was a lot of time has "flown by."

Astronauts can't afford to take time for granted. Time is integral to all that we do. We only have so much of it allotted to us while in orbit and every second is precious. In space-related operations, time is measured with respect to several different references. There is mission elapsed time (MET), phase elapsed time (PET), Greenwich Mean Time (GMT, also called Zulu time), and of course local time. (GMT is now officially called Coordinated Universal Time – abbreviated as UTC, not CUT – but the term GMT is still generally used worldwide.) These time

> Would this mean that . . . I am younger than I would have been had I not flown in space?

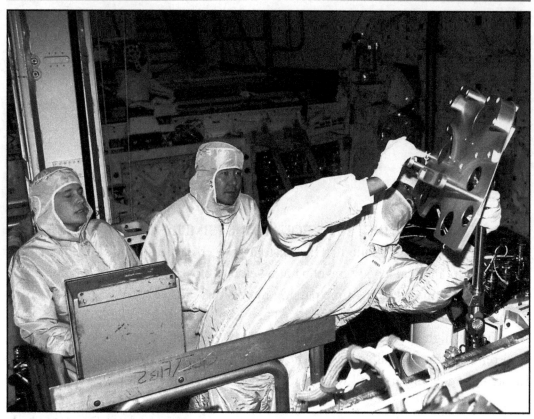

Mission Specialist Takao Doi looks on as Mission Specialist Winston Scott gets a hands-on look at some of the equipment that they will be working with during the STS-87 extravehicular activity (EVA).

standards are not at all difficult to understand, and astronauts, pilots, and people in other related professions readily communicate with one another without confusion.

MET refers to the amount of time that has passed since the launch that begins a mission. For the space shuttle, it begins at lift-off from Kennedy Space Center in Florida. For example, an MET time of 02:08:30:34 refers to flight day two, eight hours, thirty minutes, and thirty-four seconds after lift-off. It takes approximately eight and a half minutes for the space shuttle to go from the Earth's surface to orbit. Therefore, we reach orbit at approximately MET = 00:00:08:30. All activities performed in space are coordinated with Mission Control in Houston, Texas, therefore synchronized MET clocks must be maintained on board the space shuttle and in Mission Control.

PET refers to the amount of time that has passed since the beginning of a particular phase of a flight, such as extra-vehicular activity (EVA or space walk). A specific EVA task that is scheduled to begin at a PET of 02:08:30 is to start at two hours, eight minutes and thirty seconds after the beginning of the EVA.

The military, law enforcement, airlines, and other government or operational agencies use the twenty-four-hour time reference. This eliminates the confusion that might occur for example between 1:00 a.m. and 1:00 p.m. or between 2:30 a.m. and 2:30 p.m. Using the twenty-four-hour clock, 1:00 p.m. becomes 1300 (thirteen hundred hours), 2:00 p.m. becomes 1400 (fourteen hundred hours), and so on. My first space flight found us rocketing from our launch pad at approximately 0240 hours EST, 2:40 a.m. local time, and the equivalent 0740Z.

Local time is the time at any given moment in a given location on Earth. It is common knowledge that local time varies with location. For example, local 12:00 noon Pacific Standard Time (PST) actually occurs three hours later than local 12:00 noon Eastern Standard Time.

> For those who suffer from jet lag after a long trip on Earth, consider the effects of space lag.

The use of the GMT (UTC, Zulu) time standard allows time synchronization by many people, regardless of their locations on Earth (or in space). Anyone anywhere on Earth will understand the start time for an event set to begin at 1500Z (fifteen-hundred hours Zulu). Zulu time can be converted to local time by subtracting an appropriate number of hours. 1500Z equates to 11:00 a.m. EST during daylight savings time. Eastern Daylight time is Zulu minus five hours (Zulu minus four hours when EST daylight savings time is in effect).

Conventional time somewhat loses its meaning in orbit. Because astronauts in orbit see the Sun rise and set every forty-five minutes, sleep and wake cycles are programmed into our daily timeline rather than referenced to light and dark periods. The sleep / wake periods of a space mission often do not coincide with the corresponding periods at home on Earth. For those who suffer from jet lag after a long trip on Earth, consider the effects of space lag! Astronauts returning home from space typically face a period of readjustment of their body clocks to an Earth-normal cycle.

All but the most basic of activities in space appear on the crew's timeline schedule. When to eat, when to sleep, when to perform a particular experiment, when to maneuver the space shuttle, are all written into the mission timeline prior

to commencing flight. Astronauts in space must live as closely as possible to their scheduled timelines. This method is the only way astronauts can accomplish all that is required of them during a typical space mission.

Unbeknownst to them or me, my family probably began preparing me for NASA's space program long before I applied for it. Our high regard for time and timeliness began generations ago. Great Grandmother Laura Thomas is a legend within our family for her promptness. Grandmother Laura lived on a farm in rural Quitman, Georgia. I was too young to remember Grandma Laura, but in later years, after her death, my older sister LaVerne, my younger brother James (Junior), and I would spend summers on this farm visiting with our paternal grandparents, William "Papa" Scott and Grandma Eddie Mae Scott. We city kids found great fun playing in the dirt roads and exploring the woods surrounding Grandma's house. Over the years, I've heard many stories about how Grandma Laura occasionally traveled by train to surrounding towns, usually to visit with a relative or perhaps to attend a funeral. On those occasions she was insistent on leaving home early enough to arrive at the railroad station several hours before the scheduled departure of the train. It took monumental efforts on the part of other family members to restrain her to more reasonable departure times. Grandma Laura must have passed her reverence for time on to her succeeding generation. And Mama Eddie Mae and Papa undoubtedly passed it to my father, Alston. My dad was very proper, dignified, business-like, and prompt at all times.

Sunday mornings, without failure, would find our family inside the Macedonia Baptist Church in our Miami neighborhood of Coconut Grove. It took an event of monumental proportions to cause us to miss Sunday school and church on Sunday mornings. Macedonia was a large (or so we thought) church building. Huge, painted white, with a tall steeple, absent a bell, Macedonia was a place of inspiration, learning, and fun for the grownups and kids who attended there. I was a young boy when Dad became a deacon in the church and my earliest memories are of him as superintendent of the Sunday school. His respect for time was evident by the way he convened the weekly Sunday school services at Macedonia.

Opposite page – A payload canister, seen here half-open, containing the primary payloads for the STS-87 mission, is moved into the Payload Changeout Room at Pad 39B at Kennedy Space Center. The STS-87 payload includes the United States Microgravity Payload-4 (USMP-4), seen here on two Multi-Purpose Experiment Support Structures in the center of the photo, and Spartan-201, wrapped in a protective covering directly above the USMP-4 experiments.

If Saturday midnight could be considered the beginning of a mission, then at MET = 00:09:30:00 our family of five would be in the ground floor room of Macedonia where Dad would begin Sunday school. He began the opening ceremonies regardless of how many (or how few) people were present. Most often there were a few souls besides our family present at 0930. The remaining members would drift into the building beginning a few minutes after "lift-off," until twenty to thirty minutes later, when we would have a full house. By that time devotionals would be over and we would separate ourselves into our individual class groups for lesson study.

I wonder if that equates to one less gray hair?

I will never forget the confusion that erupted among some of the older members when the practice of daylight savings time was first introduced. Many of these members thought that they were supposed to set their clocks forward or backward to begin or end daylight savings time then compensate for this change in clock setting and arrive for Sunday school at the same hour as they previously had been arriving. I remember more than one person asking at the end of the Sunday school period prior to the week of the required clock change, "Deacon Scott, after we set our clocks forward, do we come to Sunday school at nine-thirty according to the old time or the new time?" Daddy would preach to them to forget "new time" and "old time." He would say, "Just set your clock ahead one hour, and when your clock says nine-thirty, you should be in Sunday school." These changes to and from daylight savings time were always made before bedtime on Saturday night, and invariably a few well-intentioned people would be an hour late (or early) for Sunday school the next day. After a week, however, the adjustment was usually made and things were back to normal.

My dad instilled the attribute of timeliness into his children by his example. My mother also possessed this same quality and exhibited it to us. It is an attribute that has served us well throughout our own personal and professional lives.

My daughter, Megan, recently said to me, "Dad, time is passing so fast. Look at all that gray hair on your head. You're getting old!" I smiled at her and replied, "Yeah; you're the cause of all this gray hair!" What she didn't realize is that Einstein might say that I'm younger than I might have been had I not flown in space. Let's see, according to Einstein I am . . . h-m-m-m, one hundred eighty six thousand miles per second, divide by . . . now multiply . . . then carry the two . . . approximately 700 milliseconds younger than I would have been. Wow! I wonder if that equates to one less gray hair?

The crew of Mission STS-87 depart from the Operations and Checkout Building en route to Launch Pad 39B, where the space shuttle *Columbia* awaits liftoff on the fourth flight of the United States Microgravity Payload and the Spartan-201 deployable satellite.

— Chapter 11 —

To Walk Outside Columbia

IT WAS THE FIRST FREE WEEKEND THAT MY WIFE, MARILYN, AND I had available to us in many weeks. Unhampered by the many social events that usually filled our Saturdays and Sundays, we talked of the relaxing, "do-nothing" two days ahead of us. We're normally early risers so we awakened at sunrise, got up and dressed, worked out at the gym, then drove the five-minute one-way distance home. No sooner had we entered the house then the phone rang. Our son on the other end said, "Turn on the TV. There's something going on with the space shuttle." I wasn't overly concerned, but I grabbed the remote and hit the power button, trying to breathe easily during the few seconds it took for the picture and sound to appear.

> ### Houston had lost radio contact with the space shuttle *Columbia*.

The TV came to life in mid-sentence, with the announcer saying Houston had lost radio contact with the space shuttle *Columbia*.

I knew *Columbia* had been in space for a couple of weeks and was due back this morning. I had followed the mission, though not fervently. It's difficult, even for a former astronaut, to stay focused on the details of each space mission while facing so many other distractions. I knew the six American astronauts on board and had flown in space with one of them on *Columbia*.

I wasn't overly concerned at the announcement of lost communications. Shuttle crews train extensively for lost comm scenarios. I thought, *the crew is perfectly capable of flying* Columbia *home for a safe approach and landing even if they can't talk to mission control.*

Then I watched the TV screen in horror when I saw white contrails streaming across the blue skies over Texas. There should have been only one, but there were several. I suspected that something had gone terribly wrong, but I tried to push out of my mind the thought that I might have lost personal friends and colleagues.

I watched helplessly as the news commentators described how NASA mission controllers were frantically attempting to regain contact with *Columbia*. The number of contrails in the TV sky told me that it was futile for the controllers to continue trying.

I watched in sadness as the events of the day began to unfold and confirm what I had hoped wasn't true; the shuttle had disintegrated.

Since the loss of *Columbia* on February 1, 2003, the thoughts of my own space flight experiences aboard *Columbia* have resurfaced.

My flight on *Columbia* in November 1999 was similar in ways to this January 2003 *Columbia* flight. Both missions were sixteen days long and focused on microgravity experiments. Both mission crews examined the effects of microgravity on plant growth, explored heart and lung functions in space, made observations of the ozone in the Earth's atmosphere, and performed some 80 other experiments. Both flights featured multinational crews. Besides the obvious differences in the conclusions, my flight included two space walks, which I performed from *Columbia*'s payload bay with my space walking partner, Japanese astronaut Takao Doi.

We commonly call it walking in space, but the technical term is Extravehicular Activity or EVA. EVA is done all the time in the movies and actor Clint Eastwood performed one in the 2000 movie *Space Cowboys*. I watched in amusement as he flew all over outer space, Superman style, saving the day.

Clint used a jet pack in *Space Cowboys*. The real jet pack is called Simplified Aid For EVA Rescue, SAFER, and is used only in cases of emergency. An errant space walking astronaut, floating loose in space, could use the SAFER unit to fly back to the space shuttle.

Big Clint – and I'm a big fan – made it look easy, but in reality EVA is a complex, physically demanding and risky activity. And although it's called

"walking" you actually float, moving your body hand over hand from place to place, very slowly, using handles mounted at strategic locations in the payload bay of the space shuttle. The space suit is actually called Extravehicular Mobility Unit or EMU. It weighs approximately 350 pounds on Earth. The 350 pounds of weight disappears in space, but its mass remains. In addition to the mass of the EMU, astronauts must maneuver their own body mass plus the mass of the tools attached to the EMU. So EVA astronauts routinely move at least 600 pounds of mass using only their hands and forearms. On my *Columbia* flight, Takao and I maneuvered close to 3,000 pounds when we manually captured a malfunctioning Spartan 206 satellite.

> More seconds passed as we each thought what no one wanted to say aloud.

It was flight day four (FD-4) of shuttle mission STS-87 when Kulpana Chawla, KC, skillfully lifted Spartan out of *Columbia*'s payload bay. KC, on her first mission in space, had been designated and trained as our Remote Manipulator System (RMS, or robot arm) operator. It was fascinating to watch her use the shuttle's cameras and the targets attached to the structure of the satellite to guide the 50-foot RMS over the satellite's grapple pin. Each movement was slow, smooth and precise. There was absolutely no room for error.

Once KC had positioned the RMS end effector over the grapple fixture pin, she squeezed the trigger on her hand controller and locked the robot arm onto the satellite. Next she gently lifted Spartan up, clearing the guides, and out of the payload bay. Very slowly she lifted Spartan higher and higher. "Looking good KC," said Kevin, our commander. High above *Columbia*, Spartan gleamed in the sunlight like some giant golden-colored alien cube. The simplicity of its shape belied its complicated construction and elaborate programming.

Opposite page – The crew of the space shuttle *Columbia* STS-87 mission pose during the Terminal Countdown Demonstration Test at Kennedy Space Center, which ends with a mock launch countdown culminating in a simulated main engine cut-off. From left to right are: Mission Specialist Winston Scott; Backup Payload Specialist Yaroslav Pustovyi (National Space Agency of Ukraine); Payload Specialist Leonid Kadenyuk (also NSAU); Pilot Steven Lindsey; Commander Kevin Kregel; Mission Specialist Takao Doi, (National Space Development Agency of Japan); and Mission Specialist Kulpana Chawla.

Spartan was designed to observe and take measurements of the Sun's corona over a 48-hour time period. Our mission was to activate Spartan, deploy it, back away from it while it took its data, and then rendezvous and retrieve it at the end of its data collection phase. We took it back home to Earth at the end of our mission so that project scientists could download its data for analysis. All went well until KC released Spartan. A successful deploy would have been evidenced by Spartan performing a pirouette maneuver 45 degrees to the left, followed by similar 45 degree turn to the right of its original position. Such a rotation would have verified a properly functioning satellite attitude control system. As KC released Spartan we crowded around the rear windows of *Columbia*'s flight deck, watching and waiting for the pirouette. Seconds seemed like minutes and no pirouette maneuver occurred. More seconds passed as we each thought what no one wanted to say aloud. Finally, Kevin called Mission Control saying, "Houston, we see negative pirouette." Houston replied, "Roger that *Columbia*. We recommend you re-grapple Spartan and we'll decide what to do from there. It's possible we only need to re-initialize the software program." "Roger that Houston. We will re-grapple Spartan."

KC's hair floated upward, away from her shoulders. It flew upward even more as she changed position in her body restraint. The restraint consisted of two parallel, horizontally-mounted cushioned rods, one mounted above the other. KC could hold herself in position to work the RMS by hooking her feet around the lower rod and hold herself stationary, sitting-style, on the upper rod. Without the body restraint, she would have floated all over the cabin. KC began moving switches on her RMS panel. She adjusted the cameras she would use for the grapple of Spartan, then released the brakes and, using her hand controller, guided the arm towards Spartan. Shifting camera views, she acquired the grapple target mounted on the satellite, and rotated the end effector to align it with the grapple fixture in anticipation of a capture. As she slowly moved the RMS forward, she paused, only for a moment, and then resumed motion. The pause was just enough to set up a slight "bounce" in the arm. It didn't seem to warrant concern. These oscillations are common, usually slight in magnitude, and generally dampen out quickly. But as KC moved the RMS end effector over the Spartan grapple fixture, the oscillating caused her to miss the grapple. Realizing she had missed, KC attempted to pull the arm back, away from Spartan, to begin a new approach. But the retreating, bouncing RMS lightly impacted Spartan, tipping it and sending it into a slow spin.

The resulting leap in our collective blood pressures when we realized our plight probably sent our Houston flight surgeon half-way to orbit. There, outside *Columbia*, our three-thousand-pound, ten-million-dollar research satellite was in a spin, outside our grasp. The spin was slow; at least it appeared slow. Perhaps KC could grab it with the RMS as the grapple pin swung by. We watched and waited, calibrating our eyes to the spin velocity, anticipating the approaching grapple fixture. Then a stab, a miss, and another miss convinced us that this was not the right way to retrieve our precious cargo. We absolutely did not want a collision between Spartan and our RMS.

STS-87 Mission Specialist Takao Doi, Ph.D., of the National Space Development Agency of Japan, is assisted with his ascent and re-entry flight suit by Dave Law, USA mechanical technician, in the white room at Launch Pad 39B as Dr. Doi prepares to enter the space shuttle orbiter *Columbia* on STS-87 launch day. At right wearing glasses is Danny Wyatt, NASA quality assurance specialist.

Kevin activated Columbia's Reaction Control System (RCS) jets and pulled out on his translational hand controller so as to back shuttle away from the satellite. The RCS jets are used to very precisely and finely maneuver the space shuttle. White-hot exhausts appeared from the RCS as Kevin slowly and carefully backed us away from Spartan. "Let's watch it for a while and see what it does," he said. "When we understand its motion, we'll move in closer and grapple it." We watched, attempting to analyze Spartan's motion. "Looks stable enough," we all agreed. "Let's give it a try." Kevin pulsed his hand controller a couple of times. *Columbia* began to move in the direction of Spartan. The position of the Spartan grapple fixture told us that we needed to rotate *Columbia* to align the RMS with the satellite. Kevin initiated a slow roll of *Columbia* to align the RMS and the satellite grapple fixture. Slowly we approached Spartan, rolling the giant space shuttle to a position calculated to allow us to recover our satellite and our pride. And as we were just about in position to grapple Spartan, it was suddenly, almost supernaturally, in a different orientation, rotating in a different direction, with its grapple fixture pointing away from us. "That's incredible," we all agreed. What's happening here? Didn't we perform the correct maneuver, in the correct direction, for the correct amount of rotation? We were looking good, then suddenly . . . What's going on?

> "OK, knock it off *Columbia*. You guys are getting low on fuel.

Again we moved *Columbia* away from Spartan, analyzed the motion, and then moved in again, only to be avoided. Could Spartan be alive and toying with us? And what about our fuel? Each maneuver was costing us precious fuel. Our mission had not been planned with this type contingency in mind. No one thought we would have to chase a satellite all over creation. "How are we doing on gas, Houston?" Kevin asked. Mission Control responded, "Not bad yet *Columbia*. You can continue your rendezvous work a while longer. We'll keep an eye on your fuel schedule."

OK, let's try this again, we decided. And once again Kevin pulsed the RCS jets and we began to move toward the wayward satellite. KC was standing by, ready with the RMS, to attempt a capture once we were in position. And once again, as we flew towards Spartan, we rotated *Columbia* to what we thought was a good

Opposite page – Looking like he's playing with a high tech "soccer ball," STS-87 Mission Specialist Winston Scott reaches out and retrieves the free-flying AERCam / Sprint (Autonomous EVA Robotic Camera).

Reflections From Earth Orbit

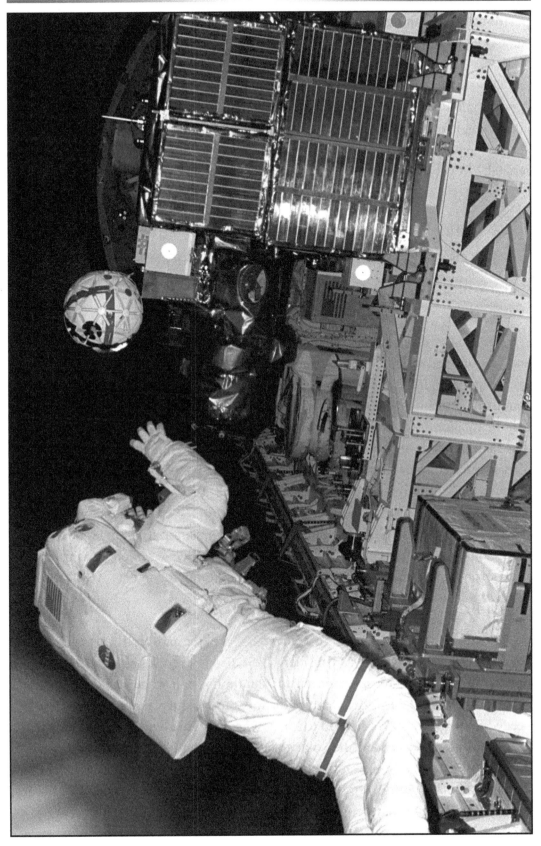

position for satellite capture. And once again Spartan suddenly was not in the same orientation as it had been only a moment before.

The sterile, monophonic sound from the flight deck speakers broke our concentration when Houston announced, "OK, knock it off *Columbia*. You guys are getting low on fuel. Why don't you back away from Spartan. We'll give you a station-keeping position and we'll take our time and figure this thing out." Kevin replied, "Roger that Houston. We'll back away and park."

> **We weren't sure that we could manually catch Spartan and then stop the spin of the three thousand pound satellite without damage to our suits and injury to ourselves.**

The next two days were spent collaborating with Mission Control. Ground analysis of Spartan's motion revealed something that wasn't apparent to the naked eye – Spartan, although rotating slowly, was exhibiting complex motion about its spin axis. Essentially, various gravitational imbalances on the satellite caused its axis of rotation to constantly shift. Instead of a fixed axis of rotation, Spartan's axis had a wobble in it – a slight one to be sure, but enough to cause problems. This is referred to as nutation, and the same things happens to planets and moons because of the multiple (and mutual) gravitational forces in the solar system. The satellite's axis was nutating (wobbling) slowly, but in the time necessary to sight on the axis and then move the shuttle and RMS into position to grapple the satellite, the axis had shifted far enough that satellite and RMS were no longer aligned. We were always a step behind the complex motion of a moving target. *Aha!* I thought. *Spartan was toying with us!*

There was much discussion in *Columbia*'s cockpit and on the ground. Every possible retrieval scenario was considered. Someone even suggested that we leave Spartan in orbit to be retrieved on a later shuttle mission when sufficient time and fuel would be available. But that option was unnecessary and unacceptable. We needed Spartan back. We needed to know what had gone wrong and how to fix it. Also, the scientists needed the solar data that Spartan was designed to acquire. There must have been reservations on the part of some people, but it was clear that the best option for retrieving Spartan was for Takao and I, during our space walk, to perform a manual capture.

Capturing a satellite by hand had been done before. It was most definitely possible but it was not a trivial affair. Spartan was similar in size and mass to a small automobile. To complicate matters, Spartan had sharp edges on parts of its exterior that could puncture a space suit glove. A punctured glove in orbit could mean disaster for a space walker. The EVA crew would have to grasp the satellite only where it was safe to do so. An additional complication was that Spartan was rotating, slowly, but rotating nevertheless. We weren't sure that we could manually catch Spartan and then stop the spin of the three thousand pound satellite without damage to our suits and injury to ourselves. Also, NASA had no choice but to consider the risks to Takao, who was a Japanese astronaut, an honored guest who hadn't signed up for manually catching an out-of-control US satellite. In fact, our government had to get permission from the Japanese government to allow Takao to participate in this risky space walk. And to further complicate the situation, Takao was on his first space flight, would be conducting his first space walk, and though he spoke excellent English, it was for him a second language.

The engineers at NASA's Johnson Space Center analyzed Spartan's systems diagrams and concluded that Spartan's Minimum Reserve Shutdown Mode should activate and the satellite should settle into a stable attitude before our attempted manual capture. This should have happened, but there was no way of knowing for sure knowing for sure if it had until we went EVA and inspected Spartan up close.

Ground support personnel were hard at work with their analysis, while astronauts and support personnel simulated manually capturing a satellite in the Neutral Buoyancy Lab (NBL), the large swimming pool-like facility used for space walk training. Veteran astronauts who had experience handling large masses in orbit were consulted during the NBL sessions to help prepare a set of recommended procedures for our use.

Because of my previous flight experience, which included EVA experience, I was designated EV-1, the lead space walker, and Takao was designated EV-2. As EV-1, I was to direct the capture process from my vantage point in *Columbia*'s payload bay. I anchored myself into foot restraints on one side of the payload bay and Takao did likewise on the opposite side. From my position, I had the "big picture" and gave Kevin directions to precisely position the shuttle so that Takao and I could reach out and catch Spartan. The manual capture consisted of catching Spartan, rotating it to a prescribed attitude, and moving it as necessary to insert it and lock it into *Columbia*'s cargo bay.

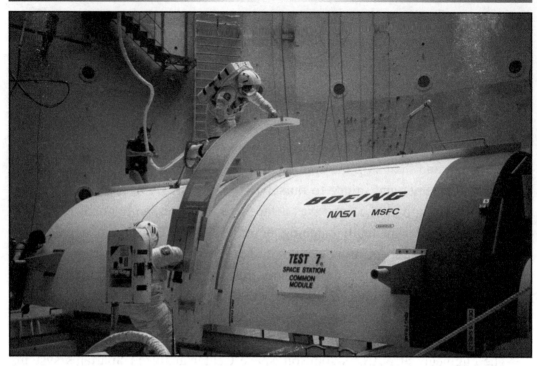

This image shows two astronauts practicing construction techniques in the Neutral Buoyancy Simulator (NBS) at Marshall Space Center (MSFC). The NBS provides a weightlessness environment for astronauts and engineers to plan and test on-orbit operations.

While the ground crews were busy with their tasks, we were equally busy in orbit. We made a toy practice Spartan from a film canister and cylindrically shaped earplugs. We used this Spartan "simulator" to practice the capture and communications protocol we were going to use to retrieve the real Spartan. We had to decide on a specific voice protocol so as to eliminate the chance of confusion between Takao and I. We decided that all motion commands would be given with reference to the payload bay. If I were to say, "Move Spartan forward," it would mean to move the satellite towards the nose of the shuttle. If I were to say, "Move Spartan up," it would mean to move the satellite upwards and out of the top of the payload bay. We all realized how important it would be for every command to be clearly understood. Maneuvering large objects in the weightlessness of space requires slow, precise, and coordinated movements to prevent loss of control and possible damage to the object being moved or the space shuttle, and to prevent injury to the astronauts.

I was pleasantly surprised at how well I rested during the sleep period before Flight Day six. I awakened at the alarm, got dressed along with the rest of the crew and, as always, ate a good breakfast. There was nervous anticipation aboard *Columbia*. We were going to perform a series of complicated and risky maneuvers that could prove disastrous. We were attempting to salvage the remainder of what had, until the loss of Spartan, been a good mission. Although somewhat anxious, we chatted and kidded with each other. We joked about having to look for work if we couldn't retrieve the satellite. We laughed hard when Kevin suggested we get Mission Control to pass to us the name and phone number of that truck driving school we often saw advertised on TV. Takao and I donned our EVA long johns and liquid cooling and ventilation garments (LCVG's) while the remainder of our crew donned their "lucky shirts." These lucky shirts, as we had humorously designated them, were the second ugliest shirts in the clothing catalog from which they were ordered. The absolutely, positively, undoubtedly ugliest shirt in the catalog was sold out and not available!

> There was nervous anticipation aboard *Columbia*. We were going to perform a series of complicated and risky maneuvers that could prove disastrous.

Commander Kevin and Steve, our pilot, in their lucky shirts, were on *Columbia*'s flight deck. They calculated rendezvous data and entered them into the rendezvous computers. They then fired *Columbia*'s rockets and we began to close the gap between us and Spartan. KC, similarly attired, assisted Takao and I through the long and tedious steps of donning and performing tests of our space suits. We checked the communications loops, cooling system, primary and secondary oxygen systems, battery voltage, suit computer, and the remainder of the systems necessary for a safe and orderly EVA. KC closed and locked us inside the airlock then assisted us via comm link with more tests. Once all of the suit tests were completed, KC took her place on the flight deck in preparation for the rendezvous and capture. Takao and I then began the complicated process of gradually depressurizing the airlock, while checking our space suit pressurization systems for proper operation. A malfunctioning space suit could spell instant death. There is absolutely no margin for error when walking in space.

Takao and I were to remain inside the airlock, on orbiter life support, until the proper moment and then switch our space suits to self-contained mode and exit the airlock, and then move to our respective locations in the payload bay. The timing of all of this had to be perfect. We knew the satellite capture would take several hours and that our space suits could only sustain us for so long on their own power. Takao and I wanted to be anchored in place in *Columbia*'s payload bay just as

Kevin and Steve were in the final stages of the rendezvous. This would give us the maximum possible length of time outside *Columbia* for the final stages of the operation – ascertaining the satellite's motion, maneuvering *Columbia* into rendezvous position, and then capturing and docking Spartan. Besides, if the Spartan capture went well, we still had a nominal EVA to conduct. The preoccupation with planning for the retrieval of Spartan made us all forget about the regular pre-planned EVA tasks awaiting us.

We configured our safety tethers, adjusted the temperature gauges on our suits, and waited.

> "Come on out Mr. Doi. It's a great day for a space walk."

When Takao and I were ready to exit the airlock, I announced to Mission Control that the EVA crew was ready and standing by. "Roger Winston," came the reply. "You are a go for EVA." Takao and I moved our suit controls to battery power and disconnected ship's power by removing our servicing and cooling umbilicals. We then maneuvered ourselves, in our awkward space suits, so that I could reach and open the hatch leading to space. I unlatched and pulled on the hatch, rotated it inward, then downward towards the "floor," and pushed it out of our way. Then I rolled over on my back and floated myself partially out of the airlock. Next I attached my safety tether to the take-up reel outside the airlock in the payload bay. I handed Takao the attachment for his take-up reel and then I floated completely outside into space. I moved upwards, slowly, towards the top of the payload bay. And as I moved to take my place on my side of the bay, I called to Takao, "Come on out Mr. Doi. It's a great day for a space walk."

The space shuttle has foot restraints mounted in its payload bay. There are retractable rod-type devices, called ingress aids, that astronauts use to aid in climbing into the foot restraints. Locking our feet in these restraints during EVA keeps us stationary in a standing-up position and allows our hands to be free for work. We take hold of the ingress aids and move our bodies as necessary to slide and lock our feet into the foot restraints. Then we release a latch and push the aid downward, out of the way, to a retracted position. We are then able to freely use our hands for work.

Opposite page – Astronaut Winston E. Scott, during one of two Extravehicular Activities (EVA) in the cargo bay of the Earth-orbiting space shuttle *Columbia*, is backdropped against a blue "blanket" of ocean water. This view was taken with a 35mm camera.

Mission Specialist Takao Doi conducts the second Extravehicular Activity (EVA) on mission STS-87. He waves at crew members inside *Columbia* from the aft Payload Bay windows.

Nothing happens in orbit exactly as it does in training. It seems that if something can go wrong, it will. As luck would have it, my ingress aid malfunctioned. It should have been easy for me to depress the ingress aid locking lever with one hand and simultaneously pull and extend the aid with the other hand. I had used the aid on my previous space flight and I had used it numerous times in training. No matter how I tried, that doggone ingress aid wouldn't budge! Finally, I shifted my position, lowered my suited body, depressed the release lever and shoved myself – and the 350 pound suit – upward. The ingress aid moved – only a few inches, but it moved. It was sticking, but not impossible for me to extend. Repeating this move, I finally got the aid extended sufficiently to allow me to get into my foot restraints and into position to catch Spartan. It was a struggle, but I managed to get it done. There was no time to dwell on it, but I wondered what was wrong with the ingress aid.

Takao had no trouble getting into position for the satellite capture. I could see him across the bay from me, waiting. I knew what he was thinking, because we both were thinking the same thing. I was rehearsing the capture in my mind, watching, waiting for Spartan to appear in my field of view.

> # The eerily quiet cube grew larger and larger in my field of view.

The Sun was well up over the Earth horizon when I leaned back and saw Spartan high above us. It required considerable effort the bend the suit backward so that I could see above my head. The bulky suit resists every movement the wearer makes. Straining, I was just able to make it comply and allow me to view Spartan. The satellite's gold-colored exterior gleamed in the brilliant sunlight as the eerily quiet cube grew larger and larger in my field of view. It looked almost like it might have been some sort of alien probe, sent to Earth to spy on us. If I hadn't known better, I might have mistaken it for an alien weapon, or perhaps a robot with strange electronic intelligence, about to attack. I knew exactly what it was, though. And I knew exactly what I needed to do. I needed to stay focused on the task at hand, direct the capture, and bring Spartan home.

Kevin and Steve flew an expert rendezvous, slowly and smoothly positioning *Columbia* so that Spartan was just above our heads. The slide of *Columbia* into position next to Spartan was cautious, almost as if we didn't want Spartan to see us coming. It was so quiet and peaceful there in orbit that I had to remind myself that we were traveling at close to eighteen thousand miles per hour. Straining, I leaned back as far as I could in the bulky space suit and watched Spartan. I had to lower my gold helmet visor because of the Sun's tremendous brightness. But I watched. Takao also watched. We all watched for a long time. I thought, *we must understand what Spartan is doing, and the current condition of its attitude control system, if we are to successfully grab it.*

The Sun rose and set twice from our perspective while we watched Spartan. With the exception of what appeared to be a minor bobble, Spartan appeared steady, with virtually no spin. The reserve attitude control mode must have kicked in and dampened the spin of the satellite. I occasionally gave Kevin position calls. "OK Kevin. Spartan is drifting a little aft in the bay. It's approximately three feet aft of me now." "Thanks Winston," Kevin replied. He pulsed the RCS jets and gently moved the shuttle to reposition Spartan directly in between Takao and I. Finally, after what seemed like a long time, Kevin said, "Hold on guys. I'm going to rotate you to a capture attitude." "Go ahead," I replied. "We're all set out here." Kevin fired the RCS jets and *Columbia* began to rotate. As the rotation began, I could see the Earth horizon begin to change out of the corner of my eye. As we rotated, the horizon, which had until now been flat and horizontal beneath us (the same as it is when you're on the Earth's surface), appeared to rotate upward towards me, giving me the illusion of falling forward. Instinctively, I moved my body in an attempt to keep the Earth flat beneath me, to keep from falling. *Vertigo*, I thought. *I've got some new weird space vertigo. I can't be distracted. I've got to direct this capture evolution!* As a pilot I was trained to fight vertigo in an aircraft by ignoring my feelings and physical sensations, and focusing on my aircraft instruments. But there had never been any training for vertigo during a space walk. In less time than it takes to relate this story, I realized that I should simply ignore my body sensations and focus on the satellite. Spartan was stable. *I'll just focus on Spartan and tune out the Earth.* It worked. As we rotated about Spartan, I felt just fine, with no sensation of falling. I again thought of the movements Takao and I would need to make to affect the capture.

> ## We had performed a successful capture, but there was more to be done.

When we had finally reached our desired orientation, Kevin stopped *Columbia*'s rotation and held us in position. "OK guys. It's all yours," he said. "Roger that," I replied. "Takao are you ready?" "I am ready," Takao replied, with diction much more precise than we natives ever used. In one final rehearsal with Takao I said, "Once more, I'm going to say stand by, stand by, capture. Do you understand?"

"I understand," Takao replied. "OK, here we go. Stand by, stand by, capture," I commanded. I put out my hands and grabbed Spartan. I was immediately struck not only by the size of the part of the satellite that I had to hold on to, but by the mass of Spartan as well. Although manipulating Spartan would be doable, manipulating it safely wouldn't be easy. "I've got my end," I said. Takao responded, "I've got my end." We had performed a successful capture, but there was more to be done.

Takao and I worked well together. I decided that Spartan needed to be rotated forward and gave the command. Takao complied and together we rotated Spartan forward to the attitude that would allow it to fit into the bay. We moved the satellite downward, towards the guides that would allow us to precisely position it in the bay to be locked down. *Something's wrong*, I thought. *Spartan is not moving completely down in its guides. Something must be blocking its movement.*

The STS-87 crew poses in their "lucky shirts" after a successful Spartan capture.

That something was my balky ingress aid. It wasn't completely stowed and was keeping us from moving Spartan far enough downward and into its berthing mechanism. An electronics box protruded from the underside of Spartan at the exact location of the ingress aid, which interfered with the docking. Now we had another predicament – I couldn't release my side of the satellite in order to further stow the ingress aid, and Takao couldn't hold the satellite alone. We certainly didn't want to release Spartan into space again while we stowed the ingress aid. As usual, we discussed options with Mission Control and we were advised to use the RMS to hold Spartan while Takao and I worked together to stow the stuck aid. Mission Control had not been sure of the status of the RMS system since the satellite malfunction, but they had decided it that was probably OK to use. Using the RMS, KC and Steve held Spartan. Takao then exited his foot restraints and moved over to my side of the payload bay, and together we were strong enough to stow the malfunctioning ingress aid. When the aid stowage was complete, Steve used the already attached RMS to place Spartan in the docking mechanism to be brought home, examined, repaired, and reflown. And there, three and a half hours after we had left *Columbia*'s airlock, Spartan was secure.

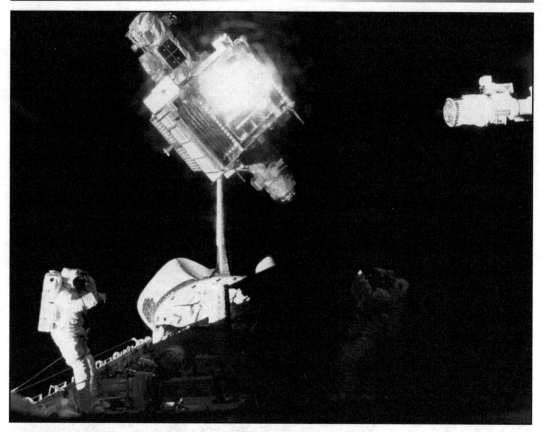
Mission Specialists Winston Scott (left) and Takao Doi prepare to recapture Spartan.

Takao and I continued on with our EVA. We tested a portable crane to be used to move large masses in space. We also tested a robotically-controlled camera with stereo vision, which could be used to inspect the outside of a space ship or station. We successfully tested tools and techniques to be used in the assembly of the International Space Station.

Nearly eight hours had passed by the time we had completed the EVA. As I was stowing our tools and safety tethers, preparing to go back into *Columbia*'s airlock, there was a tone in my headset. Glancing at my space suit computer screen I saw a thirty-minute warning telling me it was time to wrap things up because the suit was getting low on energy for life support.

No one can say why my flight aboard *Columbia* ended differently than the STS-107 mission of my friends in 2003. No one can truly understand why I am here today and the last *Columbia* crew perished. What we can say with certainty is that we must spare no effort in improving overall space shuttle safety. We must improve safety and return to flight now!

> # Nearly eight hours had passed by the time we had completed the EVA.

I remember the day I told Marilyn that I had been selected to NASA's astronaut corps. She was excited and pleased but shuddered at the thought. "That's dangerous," she said.

"There's danger in everyday living," I told her. "Nothing is 100 percent risk free. Besides, it's statistically less dangerous than the navy aircraft I've been flying all these years."

People often ask me if I would fly another space shuttle mission. I always tell them "yes." And if the setting is a bit formal I might say something like, "the benefits of space exploration far exceed the risks. I believe it is important for us to continue to reach, to learn, and to grow. It is only through exploration and growth that we are able to provide a better life for all the people of Earth." If the setting is less formal, I might add that, "flying in space is absolutely the most fantastic, incredible, exciting, wonderful, awesome, whew! . . . miraculous thing I have ever done. I wouldn't trade my experiences in space for anything!"

And as I give those two honest answers, I reflect on my childhood. I remember the hurt I felt when I couldn't build and fly model airplanes; I could only watch those who were more privileged. I remember the frustration I felt in a new and giant high school where some kids had so much and others had so little. I remember how I had always felt that, in spite of the lack of encouragement, deep down inside myself I was capable of accomplishing something bold and different with my life. I had always felt that I had the potential to rise to heights perhaps not evident by anything I had done up to that time. And as I ponder such memories, I become aware that I have exceeded any expectations that I ever had for my life. I realize again that, in those moments when I walked outside *Columbia* and directed the manual capture of a wayward satellite, I had my rendezvous with destiny and became more than I ever thought I could be.

Winston E. Scott (Captain, USN, Ret.)
Official NASA Portrait

Brief Biography of Winston E. Scott
Captain, USN, Retired
NASA Astronaut (former)

Personal Data:
Born August 6, 1950, in Miami, Florida. Married to the former Marilyn K. Robinson. They have two children. He enjoys martial arts and holds a 2nd degree black belt in Shotokan karate. He also enjoys music, and plays trumpet with a Houston-based Big Band. In addition to flying general aviation aircraft, he is an electronics hobbyist.

Education:
Graduated from Coral Gables High School, Coral Gables, Florida, in 1968; received a Bachelor of Arts degree in Music from Florida State University in 1972; a Master of Science degree in Aeronautical Engineering from the U.S. Naval Postgraduate School in 1980.

Organizations:
American Institute of Aeronautics & Astronautics; National Naval Officers Association; Naval Helicopter Association; Alpha Phi Alpha Fraternity; Phi Mu Alpha Sinfonia Fraternity; Skotokan Karate Association; Association of International Tohgi Karate-Do; Bronze Eagles Association of Texas.

Experience:
Scott entered Naval Aviation Officer Candidate School after graduation from Florida State University in December 1972. He completed flight training in fixed-wing and rotary-wing aircraft and was designated a Naval Aviator in August 1974. He then served a 4-year tour of duty with Helicopter Anti-Submarine Squadron Light Thirty Three (HSL-33) at the Naval Air Station (NAS) North Island, California, flying the SH-2F Light Airborne Multi-Purpose System (LAMPS) helicopter. In 1978 Scott was selected to attend the Naval Postgraduate School at Monterey, California, where he earned his Master of Science degree in Aeronautical Engineering with avionics. After completing jet training in the TA-4J Skyhawk, Scott served a tour of duty with Fighter Squadron Eighty Four (VF-84) at NAS Oceana, Virginia, flying the F-14 Tomcat. In June 1986 Scott was designated an Aerospace Engineering Duty Officer. He served as a production test pilot at the Naval Aviation Depot, NAS Jacksonville, Florida, flying the F/A-18 Hornet and the A-7 Corsair aircraft. He was also assigned as Director of the Product Support (engineering) Department. He was next assigned as the Deputy Director of the Tactical Aircraft Systems Department at the Naval Air Development Center at Warminster, Pennsylvania. As a research and development project pilot, he flew the F-14, F/A-18 and A-7 aircraft. Scott has accumulated more than 4,000 hours of flight time in 20 different military and civilian aircraft, and more than 200 shipboard landings. Additionally, Scott was an associate instructor of electrical engineering at Florida A&M University and Florida Community College at Jacksonville, Florida.

NASA Space Flight Experience:
Scott was selected by NASA in March 1992, and reported to the Johnson Space Center in August 1992. He served as a mission specialist on STS-72 in 1996 and STS-87 in 1997, and has logged a total of 24 days, 14 hours and 34 minutes in space, including 3 spacewalks totaling 19 hours and 26 minutes.

NASA Space Flight Experience (continued):

STS-72 *Endeavour* (January 11-20, 1996) was a 9-day flight during which the crew retrieved the Space Flyer Unit satellite (launched from Japan 10-months earlier), deployed and retrieved the OAST-Flyer satellite, and conducted two spacewalks to demonstrate and evaluate techniques to be used in the assembly of the International Space Station. The mission was accomplished in 142 orbits of the Earth, traveling 3.7 million miles, and logged him a total of 214 hours and 41 seconds in space, including his first EVA of 6 hours and 53 minutes.

STS-87 (November 19 to December 5, 1997) was the fourth U.S Microgravity Payload flight, and focused on experiments designed to study how the weightless environment of space affects various physical processes, and on observations of the Sun's outer atmospheric layers. Scott performed two spacewalks. The first, a 7-hour, 43-minute EVA featured the manual capture of a Spartan satellite, in addition to testing EVA tools and procedures for future Space Station assembly. The second spacewalk lasted 5 hours and also featured space station assembly tests. The mission was accomplished in 252 Earth orbits, traveling 6.5 million miles in 376 hours and 34 minutes.

Winston Scott retired from NASA and the U.S. Navy at the end of July 1999 to accept a position at his alma mater, Florida State University, as Vice President for Student Affairs.

(December 1999)

Winston Scott is currently Executive Director of the Florida Space Authority, where he is responsible for the statewide development of space-related industrial, economic and educational initiatives. He represents the State's interests in the development of space policies and programs and advises the Governor and Lieutenant Governor on all civil, commercial and military space matters.

Prior to his selection to Florida Space Authority, Scott was a Professor with the Florida Agriculture and Mechanical University (FAMU) and Florida State University (FSU) College of Engineering. He previously served as Vice President for Student Affairs at FSU from 2000 until 2003. He came to FSU as Associate Vice President with the Division of Student Affairs in 1999.

Mr. Scott also served as an Associate Instructor, Electrical Engineering at Florida Community College, Jacksonville, Florida, and Florida A&M University, Tallahassee, Florida.

Winston E. Scott is a FSU Omicron Delta Kappa "Grad Made Good" recipient and has received the NASA Space Flight Medal (two), Defense Superior Service Medal, Defense Meritorious Service Medal, National Defense Service Medal (two), Expert Rifleman Medal, Sharpshooter Pistol Medal, Navy "E" Ribbon (two) and the Sea Service Deployment Ribbon.

(May 2004)